T0093025

Handbook of
Optical Interconnects

OPTICAL ENGINEERING

Founding Editor
Brian J. Thompson
University of Rochester
Rochester, New York

Handbook of
Optical Interconnects

edited by
Shigeru Kawai
University of Industrial Technology
Sagamihara, Kanagawa, Japan

CRC Press is an imprint of the
Taylor & Francis Group, an **informa** business
A TAYLOR & FRANCIS BOOK

CRC Press
Taylor & Francis Group
6000 Broken Sound Parkway NW, Suite 300
Boca Raton, FL 33487-2742

First issued in paperback 2019

© 2005 by Taylor & Francis Group, LLC
CRC Press is an imprint of Taylor & Francis Group, an Informa business

No claim to original U.S. Government works

ISBN-13: 978-0-8247-2441-2 (hbk)
ISBN-13: 978-0-367-39288-8 (pbk)

Library of Congress Card Number 2004061862

Library of Congress Cataloging-in-Publication Data

Handbook of optical interconnects / edited by Shigeru Kawai.
      p. cm. – (Optical engineering ; v. 100)
  Includes bibliographical references and index.
  ISBN 0-8247-2441-0 (alk. paper)
  1. Optical interconnects—Handbooks, manuals, etc. I. Kawai, Shigeru, 1958- II. Optical engineering (Marcel Dekker, Inc.) ; v. 100.

TA1660.H36 2005
621.382'7—dc22   2004061862

Visit the Taylor & Francis Web site at
http://www.taylorandfrancis.com

and the CRC Press Web site at
http://www.crcpress.com

Preface

Rapid progress in the Internet has led the explosive spread of the Ethernet, which is its protocol on the data-link layer. Optical interconnection technologies are widely used in communication speeds over 1 Gb/sec (1000BASE, Gigabit Ethernet). Furthermore, in the next generation of 10 Gb/sec (10GBASE), they are certain to become central. These technologies are utilized in other fields of optical interconnections, such as the IEEE 1394 standard, Fiber Channel, asynchronous transfer mode (ATM) communications, and so on. Optical fiber communications have become familiar because of progress in FTTP (fiber to the premises) in the United States or FTTH (fiber to the home) in Japan. Ethernet technology is introduced in these access networks using FTTP or FTTH. This means that the distance between optical interconnection and optical communication technologies is shortening. In 10GBASE, the standard for a distance of 40 km is supported. Optical interconnection technologies may spread to metro networks.

Important points in optical interconnects are cost and module size. Low-cost, small-size technologies are required when compared to optical communications. For example, vertical cavity surface-emitting lasers (VCSELs) have garnered attention because of their ability for on-chip tests. They are used in the fields of local-area networks (LANs) by using short-wavelength devices and wide-area networks (WANs) by using long-wavelength devices. Plastic optical fibers (POFs) have merit in their manufacturing cost. They are used in the field of short-range LANs. Coarse wavelength division multiplexing (CWDM) technologies with low cost will be introduced in this field. To obtain low-cost, small-size modules, *jisso* technologies are very important. Jisso is a Japanese term that includes alignment, assembly, mounting, and packaging.

This book includes technologies from devices to systems on optical interconnects. Various low-cost, small-size technologies are discussed. More than 300 figures assist in explaining the technologies. Each author plays an active part globally in this field, and most belong to a world-famous leading company.

Shigeru Kawai

Contributors

Kenjiro Hamanaka
Nippon Sheet Glass
   Co. Ltd.
Japan

Hank Hashizume
NSG America Inc.
United States

Takehiro Hayashi
Tycoelectronics AMP K.K.
Japan

Nobuo Hori
Topcon Corporation
Japan

Takaaki Ishigure
Keio University
Japan

Takeshi Kamimura
Fuji Xerox Co. Ltd.
Japan

Takeo Kaneko
Seiko-Epson Corporation
Japan

Kashiko Kodate
Japan Women's University
Japan

Seiji Koizumi
Fusion Knowledge Network
   Co. Ltd.
Japan

Takaaki Miyashita
Ricoh Co. Ltd.
Japan

Takeshi Nakamura
Fuji Xerox Co. Ltd.
Japan

Hiroaki Nishi
Keio University
Japan

Shinji Nishimura
Hitachi Ltd.
Japan

Hironori Sasaki
Oki Electric Co. Ltd.
Japan

Naoya Uchida
Furukawa Electric
  Co. Ltd.
Japan

Osamu Ueno
Fuji Xerox Co. Ltd.
Japan

Hidenori Yamada
Fuji Xerox Co. Ltd.
Japan

Contents

1

VCSEL: Vertical Cavity Surface-Emitting Laser

TAKEO KANEKO

CONTENTS

1.1 VERTICAL CAVITY SURFACE-EMITTING LASER FOR OPTICAL INTERCONNECTIONS

After the explosive penetration of the Internet, every personal computer is now connected to one another over land
and sea. The amount of data on the Web has increased every
year. The number of computers and servers connected with
the Web has increased. The amount of data in a file has also
increased because large amounts of data can be handled by
the increased execution speed of the central processing unit
(CPU). First, only text-based data were used in computers.
Then, it became possible to write reports using colorful graphs
and pictures. Now, digital still cameras take photographs
that are sent to a friend via e-mail and printed by inkjet printers. Short motion pictures can also be viewed on many Web
sites.

Color printers and liquid crystal projectors (LCDs) bring
photo-grade pictures to people in homes and offices, but the
grade of the photo, image, and movie needs to be increased to
add beauty and color to human life. Soon, broadcasting pro-
grams will be watched on high-definition TV. Output devices
such as display devices and printers need to grow to display
more beautiful images. The connection technology for data must
also grow so that more data can be sent per second.

Optical data connection technology should become more
important in the future because the electrical interconnection

is limited by speed and difficulty of design. The range of optical interconnection is wide. The connection from equipment to equipment, from board to board in equipment, and from chip to chip in a board will become more optical in the future. However, optical interconnections also present problems, mainly the cost and the size of devices needed for the optical interconnection. The vertical cavity surface-emitting laser (VCSEL) is the most promising light source for optical interconnection because VCSEL has many advantages that reduce the cost of the optical interconnection system.

In this chapter, I will review the current status of VCSEL developments and introduce some VCSEL studies done by the Seiko-Epson Corporation.

1.2 THE EVOLUTION OF THE VERTICAL CAVITY SURFACE-EMITTING LASER

In 1979, the VCSEL was suggested first by Professor K. Iga.[1,2] At that time, many kinds of VCSELs were developed, mainly in the United States, Germany, and Japan. First, VCSELs with wavelengths of 850 and 980 nm were vigorously studied.[3-5] This was because high reflectivity (over 99%) could easily be obtained from a distributed Bragg reflector (DBR) using $\lambda/4$ stacks of GaAs and AlGaAs layers in both the 850- and 980-nm VCSELs, which could be fabricated on a GaAs substrate. The optical output power of 980-nm VCSELs always exceeded 850-nm VCSELs. Because the GaAs substrate is transparent to the 980-nm light and absorbs the 850-nm light, 980-nm VCSELs emit the light through the GaAs substrate and 850-nm VCSELs emit the light from the top surface of the VCSEL. This backside emission of the 980-nm VCSEL provided a large advantage for the high output power because the metal electrode on the upper DBR of a 980-nm VCSEL was able to cover the entire surface of the DBR, and the current injection to the center of the cavity of the VCSEL was efficient. In the case of the 850-nm VCSEL, the ring-shaped top electrodes were needed for light emission from the top surface. It was important to concentrate the injection current to the center of the cavity of the VCSEL.

Two network standards, Fiber Channel and IEEE 802.3 (Ethernet), adopted the 850-nm VCSEL as a light source for short-distance data transmission in 1997. The light-emitting diode (LED) in the near-infrared region had been used until then for the transmission in the short reach, but the transmission capacity of LEDs was not enough for Fiber Channel and Ethernet. Fiber Channel is a standard for the storage-area network (SAN), and Ethernet is the most popular standard for the local-area network (LAN). Both standards needed a transmission capacity over 1 Gb/sec. In 1996, Honeywell became the first company to commercialize 850-nm VCSELs.[6,7] Some transceiver makers also released 850-nm transceivers using their own VCSELs.[8–10] The optical transceivers using VCSELs are now widely used in these enterprise networks. Many manufacturers[11–15] supply 850-nm VCSELs to the market.

VCSELs with wavelengths of 1310 and 1550 nm also have long been developed. Because traditional telecommunications mostly used these wavelength bands, the realization of VCSELs with 1310- and 1550-nm wavelengths was long expected. However, the long-wavelength VCSELs had been delayed against the 850-nm VCSEL because there is no good combination of materials to make high-reflective DBR on the InP substrate.

First, a wafer fusion technique was adopted to realize long-wavelength VCSELs.[16] The active layers grown on the InP substrate were stuck on AlGaAs DBR using the wafer fusion technique. A new quaternary alloy, GaInNAs,[17] opened the door to the new stage of 1310-nm VCSELs.[18] The GaInNAs crystal is able to grow on the GaAs substrate and emits light with the 1310-nm wavelength band. The structure of GaInNAs VCSELs is not so different from the structure of VCSELs with GaAs or InGaAs active layers. The performance of GaInNAs VCSELs at Seiko-Epson is presented in section 1.3.4.

In VCSELs at the 1550-nm band, wavelength-tunable VCSELs[19,20] were proposed for the dense wavelength division multiplexing (DWDM) market. An air gap is inserted between the active layer and the upper DBR in the tunable VCSEL. The wavelength can be widely tuned by controlling the location of the upper DBR.

The transmission characteristics of plastic optical fiber (POF) have been improved.[21] The POF is expected to reduce the cost of fiber communication because of its relatively large core. The transmission window of POF using polymethylmethacrylate (PMMA) core material is at 650 nm. VCSELs with a 650-nm wavelength have also received attention as a light source for high-speed operation over PMMA-POF.[22] The 650-nm VCSELs have been eagerly studied but have difficulty operating at high temperature.[23] The low carrier confinement of the GaInP quantum well in the 650-nm VCSEL, because of the small band gap difference between the well and the AlGaInP barrier, causes a large reduction of the output optical power at high temperature.

POF using a perfluorinated polymer has also been commercialized.[24] This perfluorinated polymer shows low attenuation with the long wavelength range between 85 and 1310 nm. Making good use of the advantage of the large core size of POF and the established characteristics of 850-nm VCSELs, the combination of perfluorinated POF and 850-nm VCSEL for low-cost data communications has been reported.[25]

1.3 THE CHARACTERISTICS OF VERTICAL CAVITY SURFACE-EMITTING LASERS

1.3.1 Advantages of VCSELs

Laser diodes have been used in large quantities as the light source of the optical pickup for compact discs (CDs) and digital video discs (DVDs) in addition to use in the optical communication field. Laser diodes consist of an active layer, which generates a light, and a pair of mirrors for optical feedback. Ordinarily, the laser diode uses its facets for the mirrors. These facets are obtained by cutting the wafer to divide a chip, so that the laser diode emits light from the facets parallel to the surface of the wafer.

The VCSEL's pair of mirrors is fabricated by the epitaxial growth technique, such as a molecular beam epitaxy (MBE) or a metal-organic vapor phase epitaxy (MOVPE), parallel to the surface. Figure 1.1 shows the schematic representation of the difference between a VCSEL and an ordinal edge-emitting

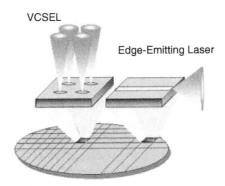

Figure 1.1   Schematic representation of the difference between a VCSEL and an ordinal edge-emitting laser.

laser. The light of the VCSEL is emitted perpendicular to the surface of the wafer. Therefore, one-dimensional (1D) or two-dimensional (2D) arrays of VCSELs can be easily fabricated. The laser light from the VCSEL is perfectly circular, and the circular beam fits the circular core of optical fibers. VCSELs can be tested in the wafer form, and a good die can be detected before the dicing process. The die bonding of the VCSEL is as easy as that of an LED. The volume of the cavity of the VCSEL is ten times smaller than that of a conventional laser. The threshold current for laser action is very low and can be under 1 mA. The small volume of the cavity is also favorable for high-speed operation. Direct modulation over 10 GHz of the amplitude of light was easily achieved in VCSELs. These advantages are favorable for optical interconnection. Optical communication systems in a shorter distance need to be less expensive and able to send more data than electrical data communication systems.

It has been speculated that the cost of VCSELs could be extremely lower than the cost of traditional laser diodes, but this expectation is questionable. Edge-emitting lasers can also be made at a very low cost. The edge-emitting lasers are used for more than 100 million pieces in the optical data storage field and sold at an even lower price than VCSELs. The cost of a semiconductor device mainly depends on the volume of

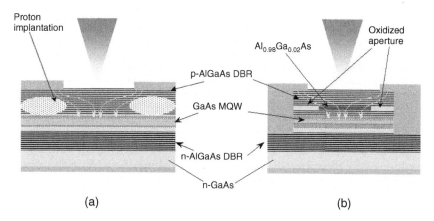

Figure 1.2 Schematic representations of the two types of VCSEL structures: (a) proton implant VCSEL; (b) oxide VCSEL.

the production of the device because the expenditure of capital for the production line dominates the cost of the device. However, there is the possibility that the cost of laser modules, such as the optical transceiver, might be reduced largely by using VCSEL and VCSEL arrays.

1.3.2   Structures of VCSELs

Two types of VCSELs have been commercialized. Their difference comes from the structure to concentrate the current in the center of the cavity. Schematic representations of the two types of VCSEL structures are shown in Figure 1.2. To form the outer region, a high-resistive proton implantation is first used (Figure 1.2a). The fabrication process of this implant VCSEL is simple, but the size control by the proton implantation makes it difficult to realize smaller VCSELs. An oxide VCSEL has an AlOx region in the outer portion of a layer of the upper DBR (Figure 1.2b). The oxide VCSEL structure is more favorable for fabrication of smaller VCSELs than the implant VCSEL. At Seiko-Epson, only the oxide VCSEL has been developed for applications to optical interconnects with high transmission capacity.

Seiko-Epson has also developed the oxide-type VCSEL with an 850-nm wavelength. The 850-nm VCSEL was grown

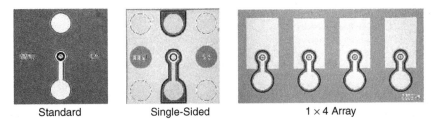

Standard   Single-Sided   1 × 4 Array

Figure 1.3 Top view of three types of 850-nm VCSEL chips: left, standard VCSEL; center, single-sided electrode (SSE) VCSEL; right, 1 × 4 VCSEL array chip.

by metal–organic vapor phase epitaxy on an n-type GaAs wafer. The active layer consists of three quantum wells of GaAs. The active layer is sandwiched by a pair of DBRs, which have high reflectivity (over 95%). The space between the DBRs is exactly 1 wavelength of the light, and the active layer is located on the center of the space. The lower and the upper DBRs are doped to be n type and p type, respectively. The electrical current flows through these DBRs to the active layer.

In Figure 1.3, the top views of the chips of Epson's VCSEL are shown. Three kinds of VCSEL chip were developed. The standard VCSEL has only the anode electrode on the top surface. The cathode electrode is evaporated on the whole backside surface of the VCSEL. Both of the electrodes are located on the top surface in the case of the single-sided electrode (SSE) VCSEL. This SSE VCSEL can be connected electrically by flip-chip bonding and preferentially operates at high frequency because the SSE VCSEL does not have to use the inductive wire. The right side of Figure 1.3 is the top view of a 1 × 4 array of VCSELs in which the spacing of VCSELs is set to 250 μm, which is useful for parallel optical interconnection using a fiber array.

1.3.3 The Optoelectrical Characteristics of 850-nm VCSELs

Figure 1.4 shows the characteristics of Epson's 850-nm VCSEL. The threshold current of 850-nm VCSELs at Seiko-Epson is 0.5 mA, and the threshold voltage is 1.6 V. The operation current for the optical output of 2 mW is 4.5 mA, and the

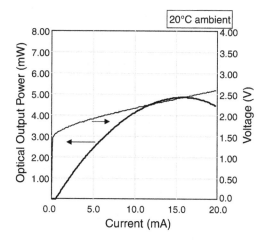

Figure 1.4 Light output power and voltage dependence on the current of Epson's 850-nm VCSEL.

wall-plug efficiency, which is the ratio of the optical power to the electrical consumption power, is as high as about 25%. Only 8 mW of the electrical power is needed for the optical output of 2 mW from the VCSEL.

The temperature dependence of light output power vs. current (I–L) characteristics is shown in Figure 1.5 in the

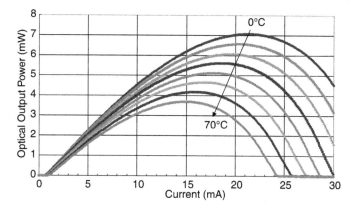

Figure 1.5 Temperature dependence of I–L characteristics of 850-nm VCSEL.

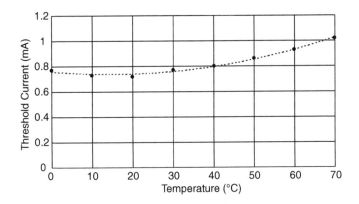

Figure 1.6   Threshold current dependence on the temperature of an 850-nm VCSEL.

temperature range from 0°C to 70°C. The dependence of I–L characteristics is quite different between conventional Fabry–Perot laser diodes (FP-LDs) and VCSELs. The threshold current of FP-LDs increases largely with rising temperature. In the case of VCSELs, the threshold current does not largely change on temperature variation because of the inherent small volume of the cavity and the small absolute value of the threshold current of VCSELs. In Figure 1.6, the threshold current dependence on the temperature of a VCSEL is shown in detail. The threshold current of the VCSEL has the minimum value at 20°C. Temperature dependence of the VCSEL is dominated by the difference of the wavelength between the gain and the longitudinal mode of the cavity. The GaAs quantum well has wavelength dependence on the temperature of about 0.3 nm/K. The temperature dependence of the longitudinal mode of the cavity of the 850-nm VCSEL is about 0.06 nm/K. The wavelength of the gain shows a red shift that is 50 times faster. The VCSEL has the minimum threshold current at the temperature at which the gain wavelength is the same as the longitudinal mode wavelength. The temperature of the minimum threshold current can be controlled by the wavelength design between the gain and the cavity.

Slope efficiency of the output power against the current decreases with the increase of the temperature in Figure 1.5.

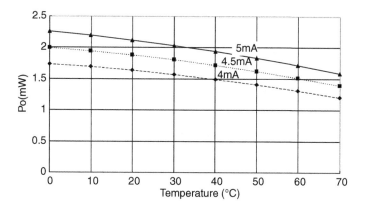

Figure 1.7 Temperature dependence of the optical output power at the constant current operation of an 850-nm VCSEL.

This decrease of the slope efficiency is larger than for the conventional Fabry–Perot laser. It is also mainly because the wavelength of the gain moves away from the mode wavelength of the cavity of the VCSEL. The wavelength design between the gain and the cavity also affects the temperature dependence of the slope efficiency. In Figure 1.7, the temperature dependence of the optical output power at the constant current operation is shown. This 2-dB reduction of the output power from 0°C to 70°C needs power monitoring in the actual usage of the VCSEL in the optical transmitters. In the case of the FP-LD, which did not show large variation of the slope efficiency, mainly the bias current of the laser was feedback controlled using the current from the monitor photodiode located at the backside of the FP-LD. Both the bias current and the modulation amplitude have to be controlled for the constant power operation of the VCSEL. Because the 850-nm VCSEL cannot emit light from the backside through GaAs substrate, the partially reflected light of the top emitted light has been monitored to control the output power from the 850-nm VCSEL. When VCSEL arrays are used in transmitters, the structure of the VCSEL transmitter for monitoring is a large problem.

Figure 1.8 shows the wavelength dependence on the temperature of the 850-nm VCSEL. As mentioned, the emission

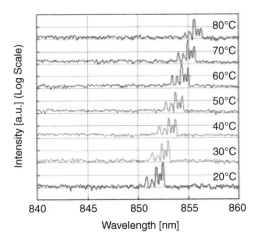

Figure 1.8 Wavelength dependence on the temperature of an 850-nm VCSEL.

wavelength of the VCSEL increases with the temperature at the rate of about 0.06 nm/K. The VCSEL is inherently the laser with a single longitudinal mode, but it is not necessary to have a single transverse mode. VCSELs with a single transverse mode have been developed, but the 850-nm VCSEL with multiple transverse modes has been commonly used for multimode fiber. VCSEL transceivers in Ethernet and Fiber Channel are designed to communicate data over multimode fiber with 50- or 62-μm core diameters. The bandwidth of the multimode fiber is influenced by the launch condition of the fiber. Especially when a single-mode laser illuminates the center of the core of multimode fiber, the bandwidth of the fiber is likely to change because only some specified modes can be excited by this launch condition.

Figure 1.9 shows the temperature dependence of I-V characteristics of the 850-nm VCSEL. The forward voltage of the VCSEL decreases slightly with the temperature. The differential resistance of the VCSEL is rather higher than that of the traditional edge-emitting laser because the current of the VCSEL flows through the high-resistive DBR layers. The differential resistance also decreases with the temperature, from 60-Ω at 0°C to 45-Ω at 70°C.

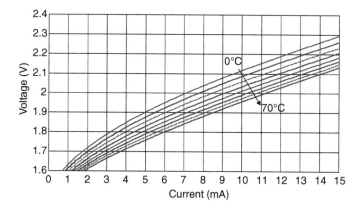

Figure 1.9 Temperature dependence of I–V characteristics of an 850-nm VCSEL.

The characteristics of the VCSEL are greatly affected by the diameters of the active region. The size of the oxide aperture controls almost all the characteristics in oxide-type VCSELs. VCSELs with shorter active region diameters have decreased threshold current and optical output power because of the less-transverse modes. Too small an oxide aperture increases the electrical resistance and affects the reliability of the VCSEL.[26]

1.3.4 Characteristics of 1310-nm VCSELs

In the 850-nm wavelength, the optical fiber made from silica has some attenuation of light, so the distance of the optical interconnection using the light with the 850-nm wavelength is limited. In the Gigabit Ethernet standard, the connection reach of the 850-nm VCSEL transceiver over multimode fiber is determined as 550-nm. The commercialization of VCSELs with a wavelength of 1310 nm is eagerly expected to reduce the cost of optical data communication in the medium-reach region. We have also developed the 1310-nm VCSEL vigorously.

A VCSEL with the active layer with three quantum wells of GaInNAs quaternary alloy was selected to develop efficiently. The GaInNAs crystal is able to grow on the GaAs substrate. The structure of the GaInNAs VCSEL is not so different from

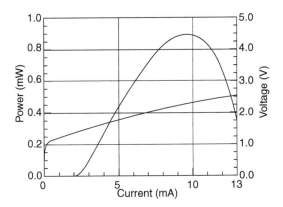

Figure 1.10 Optical output power dependence on the current of GaInNAs VCSEL with a 1265-nm wavelength.

the GaAs VCSEL except for the active layers, so the technology of the 850-nm VCSEL can be used to fabricate the 1310-nm GaInNAs VCSEL.

Figure 1.10 shows the optical power dependence on the current of the GaInNAs VCSEL. The lasing wavelength of this VCSEL was 1265-nm. The threshold current was 2.5 mA, and the maximum optical power at room temperature was 0.9 mW. The laser action of the GaInNAs VCSEL with the 1300-nm wavelength was already achieved. The present problems of the GaInNAs VCSEL are to obtain higher optical power and to certify its lifetime.

1.3.5   Reliability of 850-nm VCSELs

It was expected that VCSELs had higher reliability than conventional edge-emitting lasers because the active region of the VCSEL is not exposed in air. Honeywell has reported extensive studies about the reliability of the implanted VCSEL.[6,27] TO-CAN packaged VCSELs were life tested over 4000 h under constant direct currents from 10 to 30 mA at a temperature ranging from 100°C to 150°C. Their results clearly showed the inherent high reliability of VCSELs. According to Honeywell's reliability model, the estimated failure rate would be 0.5 ppm during 10 years of 10-mA constant current operation at 70°C.

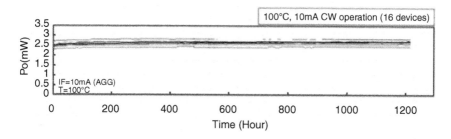

Figure 1.11 Burn-in data of Epson's oxide VCSEL at the constant current operation of 10 mA at 100°C ambient temperature.

The reliability data for oxide-type VCSELs were also reported. The very low failure rate of VCSELs causes difficulty in establishing the complete reliability model of oxide VCSELs. For their oxide VCSEL, Honeywell used the same reliability model as the implant VCSEL and estimated that the time to 1% failure would be approximately 20 years at 60°C ambient temperature and 7 mA DC. In the model, the 0.7-eV activation energy and a dependence on the square of the applied current were used.[26] Ulm Photonics reported the reliability test results for their oxide VCSEL under nonhermetic sealing. The 0.87-eV activation energy was estimated from the measured mean-time-to-failure plot against the junction temperature.[15]

In Figure 1.11, burn-in data for Epson's oxide VCSEL at the 10-mA constant current operation at 100°C ambient temperature are shown. There were 16 chips of the 850-nm VCSEL packaged in TO-46 CAN, and no previous screening was conducted in this burn-in test. This burn-in test has been continued, and there has been no significant decrease of optical power during the 2000-h operation. This means that a lifetime of over 500,000 h for our VCSEL at room temperature would be estimated using the activation energy of 0.7 eV.

1.3.6 The 10-Gbps Operation of VCSELs

At the transmission speed of data over 10-Gbps, there are many difficulties in the electrical connection. Electromagnetic interference (EMI) becomes a larger problem at higher speed

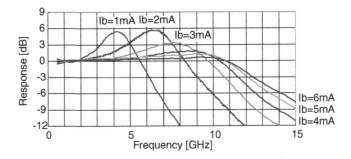

Figure 1.12 Small signal modulation response of an 850-nm VCSEL.

electrical connections. Because the electromagnetic wave does not break out from the optical fiber and the optical waveguide, the optical interconnection greatly reduces EMI problems. But, the cost of optical interconnection has been higher than that of electrical interconnection, so the optical interconnection has not been used widely. The VCSEL has the possibility of reducing the cost of optical interconnection. The high-speed operation of the VCSEL is extremely important for this purpose.

In Figure 1.12, the small signal modulation response of the 850-nm VCSEL is shown. This frequency response is the ratio of the modulation amplitude of optical power to the modulation amplitude of the electrical AC signal, which was add to the DC bias current Ib and drives the VCSEL. The response at 1 GHz is represented as 0 dB.

When the bias current is over 4 mA, the response curve is almost flat from 1 to over 10 GHz. This means this VCSEL with the bias current of over 4 mA can be used in the 10-Gb/s connection. Figure 1.13 shows the eye diagram of 10-Gb/s operation of this VCSEL at the 4-mA bias current. The eye is clearly opened, and the mask hit is not shown. This low-bias current is very important for the realization of the optical interconnection systems because the heat generation from the driver IC for the laser becomes the most troublesome problem at high-frequency operation. It is important to note that the heat generation from the VCSEL itself may not be large, but that from driver IC should be large when the drive current for the VCSEL is high.

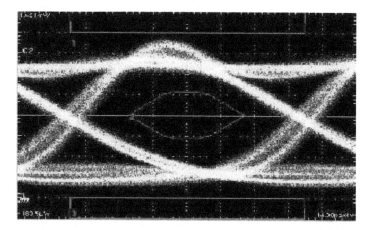

Figure 1.13 Eye diagram of 10-Gb/s back-to-back operation of an 850-nm VCSEL at the bias current of 4 mA. The extinction ratio was set to 7.7 dB.

1.4 THE MICROLENS ON VERTICAL CAVITY SURFACE-EMITTING LASERS

1.4.1 The Alignment of Optical Devices

The largest problem for getting wider applications for optical interconnections is reducing the cost of the optical transceiver. Compared with electrical interconnection, the optical system's disadvantage is the need for fine alignment. The electrical devices do not have to align as exactly with the electrical wire for their connection. The optical devices, such as lasers and photodiodes, should align with the optical fibers with an accuracy of less than several micrometers for the communication of optical signals. To reduce the cost of optical interconnection, the alignments of optical devices need to become easier.

In the optical connection using optical fiber, the light from VCSELs should enter in the core of the fiber. In the optical transceiver of Gigabit Ethernet and Fiber Channel, a transmitting optical subassembly (TOSA) is used to connect the light from the VCSEL to the fiber core. The TOSA consists of a TO-CAN package of a VCSEL, a lens, and a sleeve, in which the fiber is inserted. The light from the VCSEL is focused by

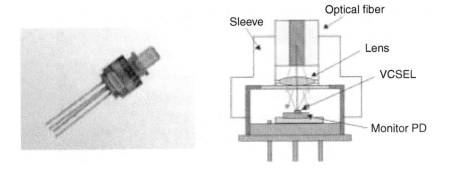

Figure 1.14   Photograph and a schematic representation of the TOSA of an 850-nm VCSEL.

the lens and enters in the fiber. In Figure 1.14, a photograph and a schematic representation of the TOSA are shown.

To align the VCSEL, lens, and fiber, the active alignment method is commonly used. In the active alignment method, the optical power from the VCSEL through the fiber is measured, and the TO-CAN is adjusted to locate the position where the measured optical power becomes the maximum value. The search for the best position is difficult, and active alignment is time-consuming. If one can make the divergence angle of the light from the VCSEL narrow enough, the positional margin of the VCSEL increases, and the VCSEL can be aligned without time-consuming adjustment.

We have developed a tiny lens on the cavity of the VCSEL and tried to decrease the divergence angle of the light from the VCSEL.

1.4.2   The Method of Fabrication of a Microlens by Inkjet on VCSELs

When a droplet is ejected onto a hydrophobic surface, a spherical surface is formed to minimize the surface energy of the droplet. This spherical surface can be used as a lens if this droplet can be put exactly on the center of a VCSEL. The accuracy of the position of a droplet by ejection from the inkjet head is only about ±15 μm and is not enough for the formation of a VCSEL lens. If the lens is not aligned with the VCSEL,

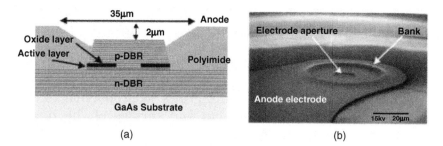

(a)   (b)

Figure 1.15 Schematic representation and the scanning electron microscopic (SEM) photograph of the bank structure for inkjet microlens formation on the cavity of an 850-nm VCSEL.

the light emitted from the VCSEL is bent by the surface of the lens.

To form a microlens exactly on the center of the cavity of a VCSEL, we fabricated a bank structure surrounding the cavity of the VCSEL. Figure 1.15 shows the schematic representation and the scanning electron microscopic (SEM) photograph of the bank structure. The bank has a 35-μm diameter, and the surface of the bank is treated to be hydrophobic. Figure 1.16

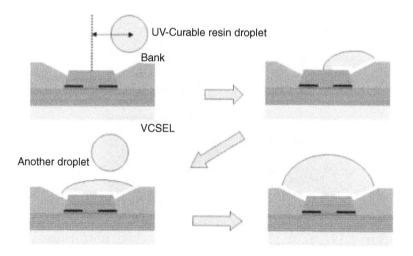

Figure 1.16 Schematic representation of the fabrication process of the inkjet microlens on a VCSEL with the bank structure.

indicates the fabrication process of the inkjet microlens on the VCSEL with the bank. Droplets of ultraviolet (UV)-curable resin are ejected from the nozzle of the inkjet head. The droplets hit on the surface and are scattered about ±15 µm but are always in the bank because the bank is large enough. The liquid in the bank grows in proportion to the number of ejected droplets, but the surface of the liquid does not go over the wall of the bank. The liquid is not able to wet the outer surface over the rim of the bank because the surface of the wall is bent abruptly at the rim. When the surface of the liquid contacts the rim of the bank, the center of the liquid automatically aligns with the center of the VCSEL. Then, the diameter of the liquid does not change, and only the top of the surface becomes higher with increased liquid volume. The radius of curvature of the liquid surface is also decreased. Afterward, the resin liquid is solidified by UV light irradiation. Figure 1.17 shows SEM photographs of the inkjet microlenses made from 3, 5, and 7 droplets, with the volume of a droplet ejected from the inkjet head set at 3.5 picoliters.

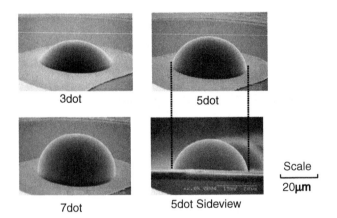

Figure 1.17 SEM photographs of the inkjet microlenses made from three, five, and seven droplets for which the volume of a droplet ejected from the inkjet head was set at 3.5 picoliters.

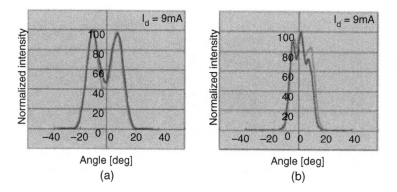

Figure 1.18 Difference of far-field pattern of the VCSEL (a) without inkjet microlens and (b) with inkjet microlens of seven droplets of UV-curable resin.

1.4.3 Inkjet Microlens VCSELs

The angle dependence of the optical power of laser light from a VCSEL without and with an inkjet microlens is shown in Figure 1.18a and Figure 1.18b, respectively. The divergence angle of the inkjet microlens VCSEL was reduced to 17° from 34° of the VCSEL without the microlens. It is also important that the transverse modes of the light emitted from the VCSEL were changed with the microlens. The output power was relatively increased, especially at the center of the angle dependence. The inkjet microlens has a possibility to control the transverse mode of the VCSEL.

To realize the easy alignment of the VCSEL without the active adjustment, the divergence angle of the light of the VCSEL should be narrower than the present value of the inkjet microlens (IJML) VCSEL.

1.5 CONCLUSIONS

The amount of data transferred in and out of the home and office has grown dramatically during the past few years, and many people are now starting to have access to the types of long-distance, large data transfers that can only be handled

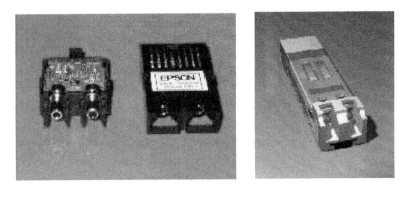

1 x 9   SFP

Figure 1.19   Optical transceivers for Gigabit Ethernet and Fiber Channel standard.

by optics. The quantity of data transmitted over networks is expected to continue to increase in the future, making optical communications technology essential for transferring data between devices. The goal of VCSEL technology is to establish a needed breakthrough for realizing the increasingly important optical interconnects between devices and between chips.

Optical transceivers with VCSELs have already been widely used in Gigabit Ethernet and Fiber Channel. Figure 1.19 shows the transceivers for these standards. These transceivers are still expensive and large for the interconnection between products in the home and for intraconnecting equipment. Demand of transceivers that use a VCSEL array and a parallel optical interface is beginning to emerge in backplane connection. These transceivers effectively decrease the required channel size of the interconnection. However, the cost of the parallel connector, such as the MTP connector, is still high and prevents the spread of parallel optical interconnection. For the evolution of optical interconnection, it is necessary to reduce the total cost of the interconnection systems considerably. VCSELs have many advantages to realize optical interconnection for data communication fields and even for consumer electronics. First, improvements of optical transceiver structure for further reduction of cost and size and improved design of VCSELs for the transceiver structure should be necessary.

ACKNOWLEDGMENTS

I acknowledge the contribution of many people in our corporate research and development division, especially Tsuyoshi Kaneko, Tetsuo Nishida, Atsushi Sato, Hitoshi Nakayama, Satoshi Kito, Mitsuru Takaya, Tsugio Ide, Shojiro Kitamura, and Teruyasu Hama. All members in Prototype Manufacturing Group 1 also contributed to this work.

REFERENCES

1. O. Soda, K. Iga, C. Kitahara, and Y. Suematsu, GaInAsP/InP surface emitting injection lasers, *Jpn. J. Appl. Phys.*, 18, 2330 (1979).

2. K. Iga, Surface-emitting laser—its birth and generation of new optoelectronics field, *IEEE J. Select. Top. Quantum Electron.*, 6, 1201 (2000).

3. F. Koyama, S. Kinoshita, and K. Iga, Room-temperature continuous wave lasing characteristics of GaAs vertical cavity surface-emitting laser, *Appl. Phys. Lett.*, 55, 221 (1989).

4. R. Jager, M. Grabherr, C. Yung, R. Michalzik, G. Reiner, B. Weigl, and K.J. Ebeling, Fifty-seven percent wallplug efficiency oxide-confined 850 nm wavelength GaAs VCSELs, *Electron. Lett.*, 33, 330 (1997).

5. J.L. Jewell, S.L. McCall, A. Scherer, H.H. Houh, N.A. Whitaker, A.C. Gossard, and J.H. English, Transverse modes, waveguide dispersion and 30-ps recovery in submicron GaAs/AlAs microresonators, *Appl. Phys. Lett.*, 55, 22 (1989).

6. J.K. Guenter, R.A. Hawthorne III, D.N. Granville, M.K. Hibbs-Brenner, and R.A. Morgan, Reliability of proton-implanted VCSELs for data communications, *Proc. SPIE*, 2683, 102 (1996).

7. Honeywell, http://content.honeywell.com/vcsel/.

8. Agilent technologies, http://www.semiconductor.agilent.com/cgi-bin/morpheus/home/home.jsp.

9. Picolight, http://picolight.com/.

10. Infineon Technologies AG, http://www.infineon.com/.

11. Emcore, http://www.emcore.com/solutions/FiberOptics/index.html.

12. Fuji Xerox, http://www.fujixerox.co.jp/product/vcsel/ (in Japanese).

13. AXT, http://www.axt.com/.

14. Truelight, http://www.truelight.com.tw/english/.

15. Ulm Photonics, http://www.ulm-photonics.de/.

16. D.I. Babic, K. Streubel, R.P. Mirin, N.M. Margalit, J.E. Bowers, E.L. Hu, D.E. Mars, Y. Long, and K. Carey, Room-temperature continuous-wave operation of 1.54-μm vertical-cavity lasers, *IEEE Photon. Technol. Lett.*, 7, 1225 (1995).

17. M. Kondow, T. Kitatani, S. Nakatsuka, M.C. Larson, K. Nakahara, Y. Yazawa, M. Okai, and K. Uomi, GaInNAs: a novel material for long-wavelength semiconductor lasers, *IEEE J. Select. Top. Quantum Electron.*, 3, 719 (1997).

18. K.D. Choquette, J.F. Klem, A.J. Fischer, O. Blum, A.A. Allerman, I.J. Fritz, S.R. Kuntz, W.G. Breiland, R. Sieg, K.M. Geib, J.W. Scott, and R.L. Naone, Room temperature continuous wave InGaAsN quantum well vertical-cavity lasers emitting at 1.3 μm, *Electron Lett.*, 36, 1388 (2000).

19. M.Y. Li, W. Yuen, G.S. Li, and C.J. Chang-Hasnain, Top-emitting micromechanical VCSEL with a 31.6 nm tuning range, *IEEE Photon. Technol. Lett.*, 10, 18 (1998).

20. C.J. Chang-Hasnain, Tunable VCSEL, *IEEE Select. Top. Quantum Electron.*, 6, 978 (2000).

21. G.–D. Khoe, T. Koonen, I. Tafur, H. van den Boom, P. van Bennekom, and A. Ng'oma, High capacity polymer optical fiber systems, *Proc. 11th Intl./Plastic Optical Fiber Conf.*, 3 (September 2002).

22. J.D. Lambkin, T. McCormack, T. Calvert, and T. Moriarty, Advanced emitters for plastic optical fibre, *Proc. 11th Intl. Plastic Optical Fiber Conf.*, 15 (September 2002).

23. M.H. Crawford, K.D. Choquette, R.J. Hickman, and K.M. Geib, Performance of selectively oxidized AlGaInP-based visible VCSELs, *OSA Trends Optics Photon. Series*, 15, 104 (1998).

24. Asahi Glass Co., Ltd., http://www.agc.co.jp/lucina/index.htm.

25. T. Kaneko, S. Kitamura, T. Ide, T. Kawase, T. Shimoda, Y. Watanabe, R. Yoshida, and Y. Takano, VCSEL module for optical data links using perfluorinated GI-POF, *Proc. 7th Intl. Plastic Optical Fiber Conf., '98,* 27 (October 2002).

26. B.M. Hawkins, R.A. Hawthorne III, J.K. Guenter, J.A. Tatum, and J.R. Biard, Reliability of various size oxide aperture VCSELs, on Honeywell's HP [7].

27. J.A. Tatum, A. Clark, J.K. Guenter, R.A. Hawthorne, and R.H. Johnson, Commercialization of Honeywell's VCSEL technology, *Proc. SPIE,* 3946, 2 (2000).

2

The Microlens

TAKAAKI MIYASHITA

CONTENTS

2.1 INTRODUCTION

The microlens and microlens arrays are widely and commonly used in measurement fields, optical communications fields, coupling with a device for optical communications and an optical fiber, three-dimensional (3D) imaging, improvement in liquid crystal projector panel light efficiency, a Shack–Hartmann sensor, and a confocal microscope. An individual microlens is formed in each sensor pixel element on image sensors such as a charge-coupled device (CCD) and a complementary metal-oxide semiconductor (C-MOS) to enhance optical light power efficiency.

The large number of numerical apertures (NAs) on a microlens provides effective receipt of the output light power from a semiconductor laser emission and acts to collimate the output light from an optical fiber to perform combined multiplexing and demultiplexing of optical switches. Furthermore, they are effective in reducing the coupling loss between optical elements.

To use the ion exchange process of glasses and plastics and the improvement of the semiconductor process for optical devices. Microlenses can be created comparatively easily from a sphere and anamorphic shapes on the lens surface in an optical system. Not only a simple single microlens but also microlens arrays are compounded and the examination for various applications is still recommended.

In connection with an improved fabrication process, the application fields have spread. Although small size microlenses are fabricated, problems in evaluation are evident. To cope with these problems, the International Organization for Standardization (ISO) provided an international standard with a terminological definition, and examination of the measuring method is still under consideration.

This chapter reports on microlens fabrication technologies, including the wafer assembly process, application of microlens arrays, and measurement of the microlens.

2.2 A DEFINITION OF THE MICROLENS AND THE MICROLENS ARRAY

Microlens and microlens arrays are defined in ISO standard 14880-1.[1]

The following is the definition of a microlens as described in the standard:

Microlens Lens in an array with an aperture of less than a few millimeters including lenses.

NOTE: The microlens can have a variety of aperture shapes: circular, hexagonal or rectangular for example. The surface of the lens can be flat, convex or concave.

Microlens Array Regular arrangement of microlenses on a single substrate.

In fact, although microlens size consideration by use differs, one of the largest sizes is for an optical pickup objective lens for an optical disk drive, and the smallest size is for the liquid crystal panel of a liquid crystal projector, with a size of about 10 μm. For optical communication applications, a lens with a diameter of about 125 μm, the diameter of an optical fiber, is also used frequently.

ISO 14880-1 defines the standard purpose as follows, with the expectation of industry activation as a widely used standard.

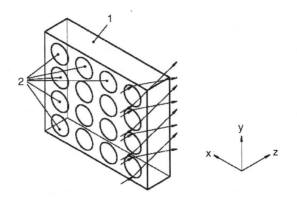

Figure 2.1   Structure of a microlens array.

[Scope] This part of *ISO 14880 defines terms for microlens arrays.* It applies to microlens arrays, which *consist of arrays of very small lenses formed inside or on one or more surfaces of a common substrate.* The *aim of this part* of ISO 14880 is *to improve the compatibility and interchangeability of lens arrays from different suppliers* and to enhance the development of technology using microlens arrays.

Figure 2.1 shows the form of a microlens (microlens array) as defined in this standard.[1] Fundamentally, although a minute lens is arranged and formed on a substrate, the microlens, which consists of a single lens, is not eliminated.

Similarly, Figure 2.2 shows the dimensions of the microlens array.[1] Especially, the definition of focal length differs from the conventional lens. In the case of a conventional lens, focal length is defined as the *distance from the principal plane to the image plane.* For the microlens, it will be difficult to use the same definition. Because the microlens normally has a large aberration for each small lens, it will be difficult to find the principal plane and image plane optimally. Also, it will be important to define the different definitions in the microlens. The distance from the vertex of the microlens to the position of the focus is given by finding the maximum of the power distribution when collimation radiation is incident

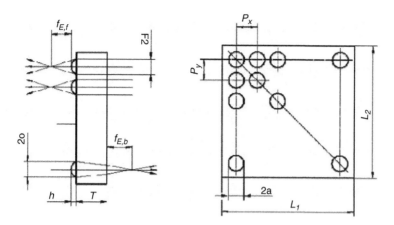

Figure 2.2 Fundamental structure of a microlens array.

from the back of the substrate, which is defined as the *effective front focal length*.

Moreover, there are three types of lens structures (Figure 2.3): the refracted-type lens (convex and concave spherical)

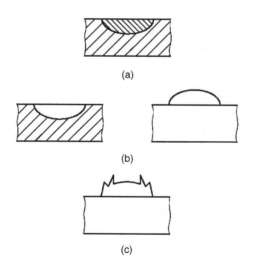

Figure 2.3 Three different types of microlens: (a) microlens with a graded refractive index; (b) surface relief refractive microlens; (c) diffractive Fresnel zone plate microlens.

TABLE 2.1 Classification of Microlens Arrays

Lens shape	Refractive (spherical) lens	Convex
		Concave
	GRIN lens	
	Diffractive lens	
Lens structure	Single lens	
	Lens array	Single-line lens array
		2D lens array

for carrying out form control, a distribution refractive index lens (gradient index [GRIN] lens), and a diffractive lens.[1]

2.3 STRUCTURE OF THE MICROLENS

There are two types of lens array structures (Table 2.1): single-lens arrays and two-dimensional (2D) lens arrays.

A representative example of the lenses classified according to Table 2.1 is shown in Figure 2.4.[2–4] Spherical, GRIN, and diffractive lenses are shown in Figure 2.4a to Figure 2.4c, respectively. A spherical lens transfers the lens shapes to the

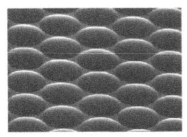

(a) Spherical lens

(b) GRIN lens

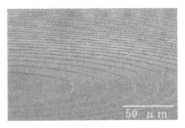

(c) Diffractive lens

Figure 2.4 Classification of a microlens array (lens structure).

glass substrate, which patterns the photoresist. A GRIN lens forms a parabolic index distribution by immersion into a salt bath made to melt low refractive index material and carries out natural diffusion of the lens mother material, which is in the shape of a cylinder and is created with high refractive index material. The refractive index is highest in the lens center because of the shape of the parabolic alignment and decreases in the outer part. The diffractive lens typically uses a chromium mask for the photoresist. This method repeats the process of exposure and development of many shapes to the substrate, creating a lattice form. Another idea is to use direct electron beam writing to develop the design pattern to the photoresist.

2.4 FABRICATION OF THE MICROLENS

The typical microlens fabrication method is shown in Table 2.2. A mechanical method of creating a direct lens with materials such as plastics and metals uses diamond or steel tools or the like. A formed metal die will be used for molding and transferring into plastic materials.

The physical and chemical methods consist of different processes. A lens can be created using photoresist formation or

TABLE 2.2   Microlens Manufacturing Methods

Method	Category	Process
Mechanical		Diamond tool
		Steel tool
Physical and chemical	Photoresist forming	Photoresist forming
		Diffractive element forming
	Photoresist and replication	Etching
	Replication	2P (photopolymer)
	Ion exchange	Ion exchange
	Direct forming	Ink jet
		Sol-gel
		Photosensitive glass
		Proton irradiation
	Direct and replication	LIGA process
Improvement	Direct forming	Step-and-repeat exposure
	Mask fabrication	Gray tone mask

diffractive element forming, which is similar to the photoresist method. Etching methods consist of forming photoresist and replication using an etching process, which is the method of transferring a photoresist shape to a substrate from the lens surface structure.

Transferring and mass-producing a lens form can be done with a 2P (photopolymer) process. The lens shape is formed by mechanical or other methods, and replication is done with a metal dye and photopolymer, assisted by UV light for transfer to a solid.

The direct forming method for a microlens uses the ion exchange process, inkjet method, sol-gel method, photosensitive glass, and proton irradiation by the synchrotron. The pattern used to generate the mask for acrylic materials is exposed, forming a 3D lens structure.

The LIGA process (from the German words for the process: Lithographie, Galvanoformung, and Abformung) uses lithography, electroplating, and molding processes to produce microstructures. Deep x-ray lithography, electroforming, and plastic molding are used.

To create an improved and complicated lens form or a diffractive lattice structure, the step-and-repeat process, which uses a moving slit pattern, changes the exposure energy of localization and creates a complicated form by multiple exposures. Gray tone mask technology offers an improved exposure patterning system. A detailed, fine pattern is written in special glass that is sensitive in the electronic beam; exposure energy concentration will be given by transmittance distribution, which controls over 1000 steps. The gray tone mask provides a high-precision surface structure for the photoresist after its development. It is applied with a suitable method or combination for every use.

Photoresist forming, diffractive element forming, etching, and gray tone mask are explained next.

2.4.1 The Photoresist-Forming Method

The photoresist-forming process is shown in Figure 2.5. A spinner is used for coating the photoresist for the substrate, and the coated photoresist thickness is derived from the viscosity

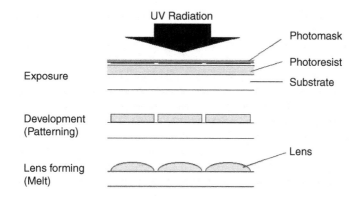

Figure 2.5   Microlens fabrication using photoresist.

of the photoresist and the rotating speed after it dries. The pattern of the chromium mask formed on the glass is used at the photoresist and is exposed by UV light. After developing, the cylinder-shaped photoresist is arranged so that a lens will remain on the glass substrate. After heating over the melting temperature of the photoresist, it will become a globular form with surface tension and is transferred from the cylinder form. The height and area of the ball-shaped lens are controlled by the type and thickness of the photoresist used.

A photoresist lens is formed completely on the substrate after cooling and drying. Figure 2.6 presents processing of the lens by the photoresist.[5]

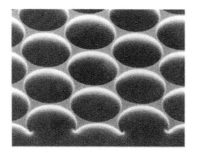

Figure 2.6   Photoresist microlens.

2.4.2 The Photoresist Forming-and-Etching Process

The photoresist forming-and-etching process is shown in Figure 2.7. A spinner is used for coating the photoresist for the substrate and the coated photoresist. The pattern of the chromium mask formed on the glass is used at the photoresist and is exposed by UV light. After developing, the cylinder shape of the photoresist is arranged so that a lens will remain on the glass substrate. After heating above the melting temperature of the photoresist, it will become a globular form with surface tension from the cylinder form. The height and area of a ball-shaped lens are controlled by the type and thickness of the photoresist used.

A photoresist lens is formed on the substrate, and then it is cooled and dried. Next, the photoresist form will transfer a 3D form such as a lens to the substrate, such as glass or another material. Generally, the dry etching process is used with activated plasma gas. Control of plasma gas and etching conditions is needed for a 3D transfer of a shape formed by the photoresist, so that so-called isotropic etching is performed in the photoresist and substrate material (lens-forming

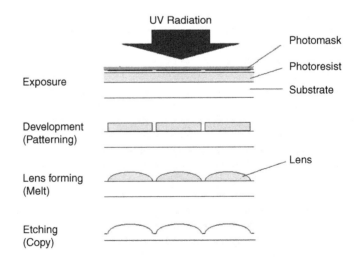

Figure 2.7 Microlens fabrication using the etching process.

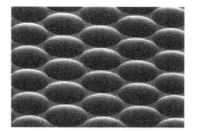

Figure 2.8 Microlens fabricated by the etching process.

material). Figure 2.8 presents an example, transferred on glass, of the lens form fabricated by the photoresist using the etching process.[6]

Etching speed depends on the mixture ratio of plasma gas. Figure 2.9 is an example of the etching speed of photoresist and GaP material.[7] The etching speed of each material is measured using the percentage of Ar gas to Ar(Cl_2) as a parameter. It is restricted when the etching speeds of the photoresist and lens formation material become the same when the composition ratio of Ar gas is as small as about 1%. Therefore, in this example, by setting up and etching the

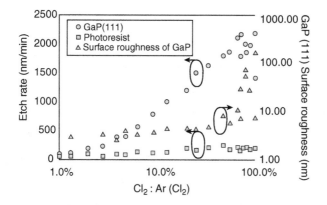

Figure 2.9 Etching rate of GaP and photoresist depends on the mixed plasma gas ratio and surface roughness quality of a GaP microlens.

composition ratio of Ar gas at nearly 1%, the photoresist form can be transferred to the GaP substrate as it is, and so-called isotropic etching can be realized. Furthermore, to realize high-speed etching, gas with the high etching rate of photoresist is chosen, or etching conditions and system changes are needed.

2.4.3 Improvement of the Photoresist Forming-and-Etching Process

The photoresist forming-and-etching method is normally used to form a two-level binary pattern on the photoresist, which is fabricated to melting. The sphere surface shape is formed because of surface tension. However, shape control is difficult, especially for nonspherical shapes, such as an aspheric surface lens. The method of putting the photoresist on in many steps and creating a brazed form in approximation is well known. For grating fabrication, when the sharp edge form is required, it would be difficult to make it incorrect.

If the surface deviation is small, the exposure value to the photoresist can be changed for each place using a dithering mask. The form requirement in approximation can be created to the photoresist. However, the vertex position of the brazed diffractive grating shape changes rapidly, or a highly precise aspheric surface lens shape deviation is also different from the spherical shape. It can be obtained to generate the shape from the design using a gray scale photomask, which changes the local concentration orientation for every place on a mask. Figure 2.10 shows generation of the gray scale photomask[8] and the microoptics element created. The three steps are as follows:

1. A pattern is recorded and developed on special photosensitive glass, which has sensitivity in an electron beam, and then a photomask is fabricated. The concentration distribution proportional to the desired form is recorded on the photosensitive glass.

2. The glass is exposed by UV light to photoresist using the photomask with which the power distribution was recorded. If negative patterns are developed, the form of photoresist with the triangle form

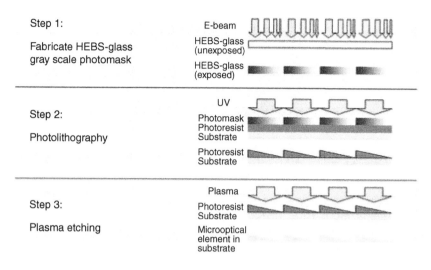

Figure 2.10   High-precision microoptics element fabrication process using gray scale photomask. (HEBS: High energy beam sensitive)

proportional to the power distribution of a photomask will remain on the glass substrate. This is melted, and the surface sphere shape is created with surface tension by the conventional method. The triangle shape is formed according to the power distribution of a photomask without the photoresist melting process, which is the method used for the gray scale photomask.

3. The 3D shape formed by the photoresist is obtained by transferring photoresist and substrate material (lens-forming material) into substrate material by isotropic etching as well as the conventional method.

Figure 2.11 is an example of the microlens array created using the gray scale photomask.[9] It is an effective means to create the lens array because of its high NA; and the aspherical shape lens occupies 100% of the fill factor for this example. Furthermore, it is also effective in creating large changes, such as the vertex part of a brazed-type diffractive grating element.

Figure 2.11   The 100% fill factor f/1 aspheric microlens arrays by
HEBS-glass gray scale lithography.

Because it has sensitivity to the electronic beam, creation
of a fine pattern is possible for the gray scale photomask
shown in the example (Figure 2.10). The use of a photosensitive glass that has sensitivity in visible and UV light, where
the beam spot size is about 0.5 μm using a laser source. The
gray scale photomask has sensitivity in an electronic beam
with a spot size about 0.1 μm, which is possible because the
electron beam is precise.

2.4.4   The 2P Replication

The method of replication for creating a microlens using a
metallic stamper and photopolymer is effective in reducing
the manufacturing cost for a microlens array. Photopolymer
is placed between the metal stamper and a glass substrate.
Then, it is exposed and fixed by UV light. Figure 2.12
represents an example of the fabrication process.[10] Conduc-
tive films first cover the original lens top. Then, the negative
lens structure shape is fabricated by electroplating to be
generated as a stamper. A photopolymer is placed between
the stamper and the clear substrate (for example, glass or
plastic); after exposure to UV light, the structure is fixed.
This is carried out by coating the cover polymer, which
further transfers the metallic mold form, and a cover glass
is added.

Figure 2.13 shows the applied lens array example of the
liquid crystal device (LCD) fabrication process.[10] The basic

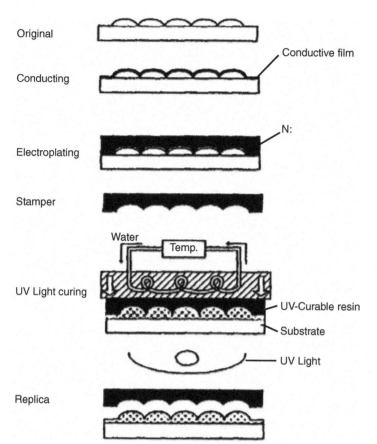

Figure 2.12 Microlens fabrication using 2P process.

process is the same as the conventional 2P process. First, a concave lens structure is formed on the glass substrate with a negative-shaped stamper. Then, the higher refractive index convex lens is formed between the concave lens shape and a cover glass. After exposure to UV light, the convex lens form is fixed between the base glass and the cover glass. When polymer material is used, the coefficients of thermal expansion are different from the glass substrate, which has a heat-resistant problem. Adding a cover glass raises reliability.

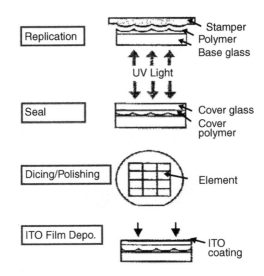

Figure 2.13 Microlens fabrication using 2P process for a LCD projection device.

2.5 MICROLENS APPLICATIONS

Application of each microlens created by the processes mentioned in the previous sections are discussed.

2.5.1 One-Dimensional Lens Array

Figure 2.14 shows a connection with the optical fiber array, which is one of the typical applications of a one-dimensional

Figure 2.14 Classification of a microlens array (single-line lens array).

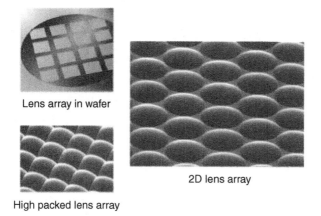

Lens array in wafer

High packed lens array

2D lens array

Figure 2.15 Classification of microlens array (2D lens array).

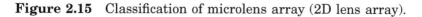

lens array.[11] This figure consists of the fiber arrays of four channels; it connects with an optical fiber array, coupling with a semiconductor laser diode array and an optical waveguide. For aspherical surface lens applications, it is made from a different curvature structure with a main axis and subaxis. It will be combined for use with an edge-emitted semiconductor laser, then it will be converted to the circular beam shape.

2.5.2 Two-Dimensional Lens Array

Figure 2.15 shows a representation of a 2D lens array.[12,13] This array is used for improved coupling efficiency of the 2D optical switch and for optical efficiency of an LCD panel of a liquid crystal projector and other such applications.

An example of image formation by the 2D microlens array is shown in Figure 2.16. Multiple different images are formed from one subject; this plurality differs by microlens array. All formed images from each lens are different. A typical application is for 3D image processing, by which restoration of a 2D picture to a 3D object is also attained.

The fabrication process of applying a 2D microlens array to an LCD panel for an LCD projector is shown in Figure 2.17.

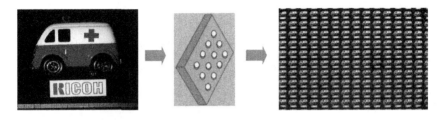

Figure 2.16   Image formation by a 2D lens arrays.

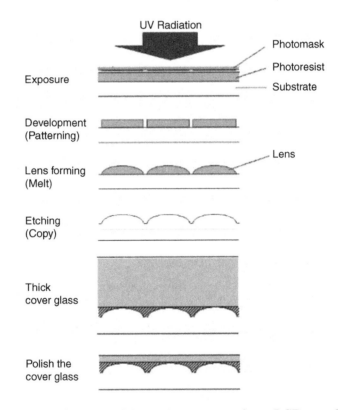

Figure 2.17   Microlens fabrication process for a LCD panel.

The microlens array has an added cover glass on the top surface of the fabricated microlens to make the completed microlens apply to the LCD panel. This is the same method used for the photoresist forming-and-etching process.

It is effective to apply the microlens fabricated surface facing the liquid crystal, requiring the surface of the microlens array to be flat. Then, the fixed surface cover glass is ground and polished. After adhesion, it forms a thin cover glass on top of the microlens array. The liquid crystal cell with a microlens array will improve optical power efficiency when the projected light passes through the thin film transistor (TFT), reducing the beam size.

2.5.3 Diffractive Device

Figure 2.18 represents color separation using a diffractive lens.[14] A diffractive lens will separate the three different colors from the incident white light: red, green, and blue. Effects such as wavelength separation and focusing from the diffractive lens can be freely provided by arbitrary diffractive lens form designs, and the application range is wide.

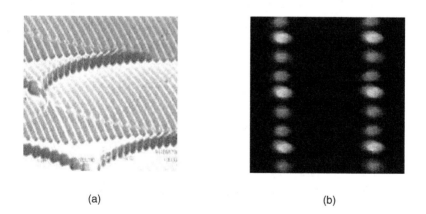

(a)   (b)

Figure 2.18 Color separation of diffractive lenses; (a) diffractive lens and (b) color separation image.

2.6 MICROLENS WAFER ASSEMBLY PROCESS

The high-NA microlens is effective in correcting the output radiation from a wide emission angle semiconductor laser and acting as the collimate of the output light from an optical fiber to perform combination multiplexing and demultiplexing of optical switches. Furthermore, it is effective for connecting the optical elements with little connection loss.

It was difficult to obtain a high-performance NA of 0.7 in previous microlenses. These consisted of the double-surface lens and a microlens several hundred micrometers in diameter. To obtain the high-NA microlens, they should consist of a three-lens surface (Figure 2.19).[15] Figure 2.19 shows the structure applied for an optical disk pickup objective lens. Using optical design software, the structure of a high-NA microlens and tolerance were calculated for the fabrication process. The calculated tolerance is in the limitation of wavefront aberration less than 0.07 ramda.

The lens diameter is 200 μm, and it has an NA of 0.85. Overall length is 708 μm. The high-NA microlens consists of three lens surfaces. L1 and L2 are convex lenses. L3 is a concave lens filled with a resin with a high refractive index. Substrates and cover glass materials are fused quartz (nD = 1.45847). The refractive index of the resin between L3 and

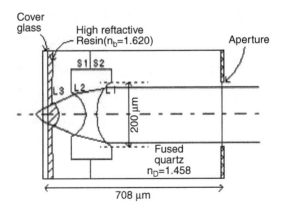

Figure 2.19   Structure of a high-NA microlens.

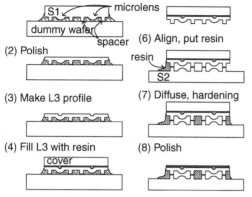

Figure 2.20 High-NA microlens fabrication process flow.

the cover glass is 1.620. The tolerances of each surface spacing and optical axis displacement are ±3 µm and ±1 µm, respectively. This structured lens achieved high NA with the 650-nm wavelength design.

Figure 2.20 shows the high-NA microlens fabrication process flow.[15] The high-NA microlens is fabricated with a wafer-assembling process. The photoresist surface structure of microlenses L1, L2, and L3 are formed by photolithography and transferred into a fused quartz substrate by dry etching. Surface spacing is adjusted by polishing or dry etching. Before this modification of process flow, the S2 substrate is polished after bonding to S1 and S2. The substrate S2 is formed with lens L2 and the dummy board, which is stuck with wax. At the polishing process, the thickness of substrate S2 was measured as the distance of the upper S1 surface to the upper S2 surface. Therefore, the measurement result was influenced by S1 surface condition. To reduce the thickness error of substrate S2, substrate S2 is polished on the flat dummy substrate, and S1 and S2 are bonded after polishing S2.

Substrates S1 and S2 and a cover glass are bonded with UV-sensitive resin. When putting substrate S1 on S2, some errors are induced by shrinkage of the resin, so the substrates are pressed strongly so they do not move after optical axis

(a)   (b)

Figure 2.21   Photograph of a high-NA microlens.

alignment. Resin is spread in the substrate gap by a capillary phenomenon. Substrates S1 and S2 are united using an alignment mark formed beforehand.

The aperture is added to the high-NA microlens. The aperture is made with Cr thin film before fabricating the high-NA microlens. The Cr film is sputtered and formed into an aperture shape by photolithography and wet etching.

Figure 2.21 is photographs of a fabricated high-NA microlens.[15] In the view from the cover glass side, the microlens is at the center, and other elements are spacers. Spacers control the L1–L2 surface spacing and prevent inflow of resin into the lens surroundings.

The white square area in the aperture side view is Cr film. The circle area at the center of the Cr film is the aperture. The surface of the aperture side was absolutely flat, so the addition of the aperture to the high-NA microlens was easy.

2.7 MICROLENS EVALUATION

2.7.1 Surface Profile Measurement

2.7.1.1 Surface Profile

2.7.1.1.1 Interferometer Measurement

The surface profile of a lens is an important factor for checking performance. There are many problems that cause inability to perform accurate measurements. The sample can be too small or the lens circumference edge too steep. Various methods are examined, and the actual condition is properly chosen.

To measure the lens surface profile, the common methods are to use stylus contact measurement equipment, a three-dimensional measuring instrument, a noncontacting system using an interferometer, and the method using a confocal microscope. The typical noncontact 3D measurement system uses a laser beam as the probe.

In early stages of measurement for the aspherical lens surface, the contact stylus-type surface profiler was used. Even the form of the stylus contact system has an error factor in the lens edge of the lens, and this method has seldom been used in the microlens surface shape measurement.

A noncontact measurement system has each feature, and the actual condition is properly used with various lens forms, sizes, and so on.

The interferometer of the Twyman–Green system is commonly used for microlens surface profile measurement. It tends to arrange the reference side near the form of the test lens surface for measurement of the surface form, although a Fizeau system is used for wavefront aberration measurement of the lens with a common interferometer. Figure 2.22 shows the Twyman–Green interferometer system.[16] This example was developed from Erlangen University in Germany. The method also was developed at National Physical Laboratory in the United Kingdom, and the product is sold by a U.S. manufacturer as a surface form of measurement for a microlens. When the interferometer of a phase shift system is applied, although the information on the whole lens side is obtained for a short time, measurement of the portion with the large slope of the lens circumference edge is difficult.

2.7.1.1.2 Laser Probe Scanning Measurement

The noncontacting 3D measurement system is commonly applied for the measurement instrument system using the laser beam as a probe. Figure 2.23 shows the measurement

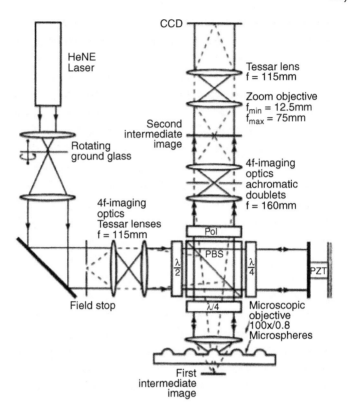

Figure 2.22   Example of the interferometer of a Twyman–Green system.

principle,[17] and Figure 2.24 shows the equipment.[18] With the laser beam focused on a sample surface using the objective microscope lens, the surface-reflected light is detected by an AF (autofocus) sensor; the entire optical sensing system is moved perpendicularly (the optical axis direction of a measurement optical system) simultaneously, and a vertical surface position is detected. The *x–y* position is scanned with a high-precision mechanical stage controlled by the optical linear control system. About 1% of the reflected light is detectable by the probe beam, and if a sample has a coarse surface, it is said that about a 90° (perpendicular) field can also be measured.

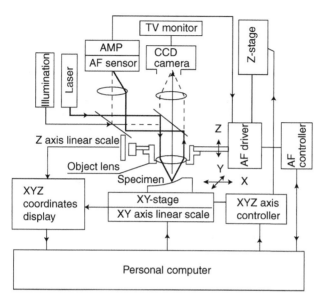

Figure 2.23 Principle chart of the laser probe microscope with a scanning stage AMP (amplifier).

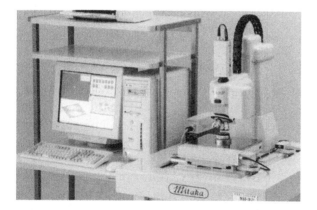

Figure 2.24 A view of the laser probe microscope with a scanning stage.

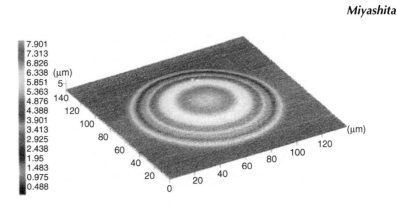

Figure 2.25 Example of the laser probe microscope with a scanning stage (sample 100-μm diameter microlens).

The whole measured surface data will be derived with a spot beam scanning measurement. It will probably become the limited use as a mass production inspection measurement from time limits.

An example of measurement of a microlens is shown in Figure 2.25. Although this is an example with comparatively easy measurement, the form on the surface of a lens is known well, and the form of the lens circumference edge especially generated by a machining error can also be grasped with high precision. Perpendicular display decomposition ability is set to a maximum of 0.001 μm.

2.7.1.2 Radius of Curvature

The radius of curvature of a lens is measured with the target image, which is set into the focal point of collimater optics and focused conventionally with a microscope object lens on the sample lens. The difference between the two positions on the optical axis (according to the lens surface spot reflection and reflection from the center of the sphere) is measured.

Figure 2.26 shows the method of improving the sensitivity using the interferometer.[19] This system is applied to the Twyman–Green interferometer system. It is also put to practical use because lens surface form accuracy may affect measurement accuracy.

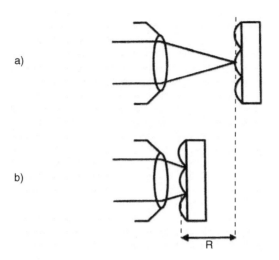

a)

b)

R

Figure 2.26 Example of curvature radius measurement using the interferometer.

2.7.2 Optical Characteristic Evaluation

2.7.2.1 Focal Length Measurement

Many test methods are used to measure the optical characteristics of a lens, which are focal length and wavefront aberration. It is difficult to apply conventional methods for microlens focal length measurement that use the principal point position by the nodal slide method because sometimes the small microlens has a large aberration, and it is difficult to find the nodal point.

The method of determining the distance from the vertex of a lens to the focal point is called determination of the effective front focal length. Incidence of the parallel light is carried to a test lens, and focal length is measured to find the position of the maximum power distribution point.

Figure 2.27 shows measurement of the focal length using the interferometer.[19] The interference pattern depends on the position of the microlens in relation to the microscope objective. If the light from the objective microscope lens is focused on a spot on the microlens surface, it is reflected in the cat's

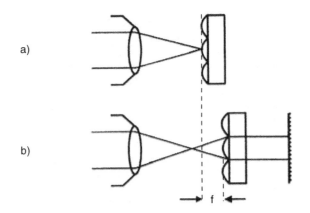

Figure 2.27 Example of focal length measurement using an interferometer.

eye configuration as shown in Figure 2.27a. Then, the micro-lens is moved from the objective lens, and the fringe pattern is changed to circular and then straight again as the configuration shown in Figure 2.27b.

At this position, the focal points of the objective lens and the microlens coincide, and the light transmitted through the microlens is recollimated and reflected by the mirror. The straight fringe is observed again on the collimated position of the microlens and objective lens. The distance between these two positions is defined as the focal length. These positions are easy to observe in the linear scale system.

By measurement light passing a test microlens twice, since a relation collapses strictly by the disturbance containing wavefront aberration of a lens, although measurement is simple, there is a fault, which is easy to include an error.

There is also another method to find a focal length without using an interference optical system using a similar optical system to detect optical intensity.

Another possibility in measurement is use of the test chart and collimater system. A chart is set into the focal position of a collimater, and a chart is projected and focused with a sample microlens to the focal plane. Then, the image size is measured, and the magnitude of the target size of the chart and image are compared and calculated. The focal length is easy to calculate from the magnitude. Sometimes, a magnification error occurs if the wavefront aberration of the sample microlens is large.

2.7.2.2 Wavefront Aberration Testing

2.7.2.2.1 Mach–Zehnder Interferometer

Many wavefront aberration measurement methods are proposed. It is best to choose the single-pass interferometer for the performance of a microlens, which sometimes has aberration problems. Figure 2.28 presents a Mach–Zehnder interferometer for a wavefront aberration measurement system for the microlens.[20] A Mach–Zehnder interferometer is a typical single-pass interferometer; it is expected to have little measurement error. It is one example of using the objective microscope lens and a test microlens together.

The wavefront aberration can be measured by the phase shift method using a parallel beam in the focal position of a test microlens on the measurement arm side. A fringe pattern is combined and generated with the sample beam and reference

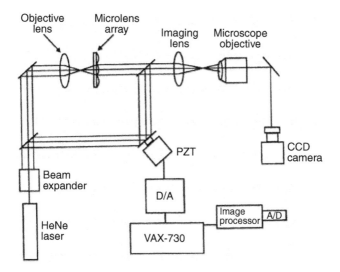

Figure 2.28 Wavefront aberration measurement using the Mach–Zehnder interferometer.

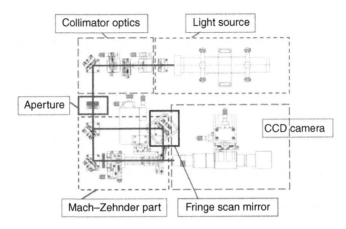

Figure 2.29 The Mach–Zehnder interferometer layout using a projection aperture.

beam on the interferometer, and it is projected on a CCD element with the wavefront of a test microlens.

Sometimes a double-pass interferometer (for example, Fizeau, Twyman–Green, and so forth) will cause a conjugate relation error. A single-pass interferometer has the advantage of lack of influence by the performance of microlens aberrations. A test microlens can be arranged on the inside arm of a measurement optical pass. Equipment composition does have restrictions.

Figure 2.29 shows the Mach–Zehnder interferometer layout using a projection aperture, which is based on an ordinary Mach–Zehnder interferometer. An effective projected aperture is added between the light source and the interferometer part.[22,23] An image of the aperture is formed on the pupil plane of the test microlens using a standard lens. The aperture's shape is projected by an objective microscope lens to the test microlens pupil plane, and an objective microscope lens focuses a parallel light to the focal position of a test microlens, which is placed in a measurement beam in interferometer optics. An effective aperture is determined in the combination of the focal length of an aperture size and an objective microscope lens. The advantage of this method is that the measurement removes the unexpected light generated by an aberration at the time of measuring the test microlens with a large aberration.

Figure 2.30 View of the Mach–Zehnder interferometer.

Figure 2.30 presents the example of a first prototype.[21–24] The piezo mirror, which is settled in the reference beam, performs a phase shift, and wavefront aberration is calculated by ordinary software. This optical system is designed in a horizontal distribution; it is easy to improve optical quality and expand an optical layout.

The example, which measures the microlens using this interferometer, is shown in Figure 2.31. A projected 18-μm

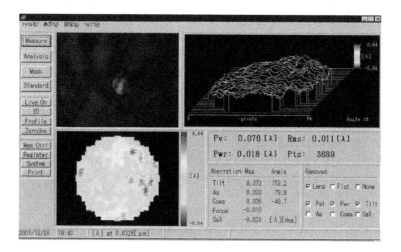

Figure 2.31 Example of a microlens wavefront aberration using the Mach–Zehnder interferometer (sample 20-μm diameter microlens).

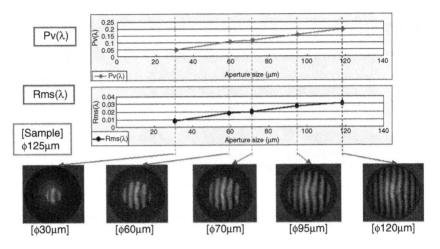

[φ30μm]   [φ60μm]   [φ70μm]   [φ95μm]   [φ120μm]

Figure 2.32 Wavefront aberration vs. diameter of the effective aperture (sample 125-μm diameter microlens).

diameter aperture size is projected on the 20-μm diameter test microlens, and wavefront aberration is measured. Comparatively good results are obtained. A peak-to-valley (PV) value of wavefront aberration is 0.076, and the RMS (root mean square) value is 0.011.

Another example measured and confirmed the effectivity of this projection aperture. The projection aperture size was from 120-μm to 30-μm in diameter in five steps on the 125-μm diameter sample microlens, and a wavefront aberration was measured as shown in Figure 2.32. It may be easy to check the effect of a projection aperture. With the lens of this example, aberration of the position area near the optical axis showed the tendency, which became small. Also, with the conventional interferometer, the measurement area can be chosen by software if data are collected by the phase shift method. If using the projection aperture, unexpected light noise will be removed physically. It especially is effective for measurement of the small-size microlens and lenses with large aberrations.

The revised and improved interferometer system is shown in Figure 2.33. This interferometer system was designed vertically. It has a large advantage in its ease of operation in testing, which means a test microlens and microlens array

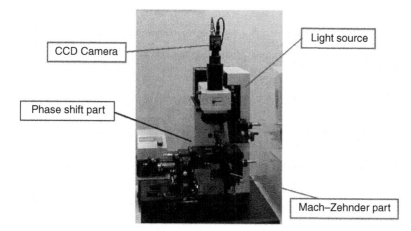

Figure 2.33 A view of the Mach–Zehnder interferometer for microlens measurement.

are settled on a horizontally operated stage. A sample stage is 6-in.² and is moved by a micrometer. The optical axis is changed with the goniometer head.

Figure 2.34 is another example of a Mach–Zehnder interferometer[25] similarly built simply with the microscope base. A laser light source beam is divided into two parts by a polarization beam splitter. The measurement and reference beams are led to a microscope by the optical fiber; the objective microscope lens is inserted in a measurement beam arm. A piezo mirror is settled in a reference beam, and phase shift data are collected by mirror shift. Two microscope objective lenses are inserted into the measurement light arm. The first is a coupling lens for the sample microlens, and the second is an objective lens for the CCD image sensor. Two objective microscope lenses arranged in a measurement optical pass in an interferometer might be considered to remove the influence from an aberration of the objective lens. This system has a simple structure, which is achieved by simple operation.

The fiber-based Mach–Zehnder interferometer system is shown in Figure 2.35. The example of measurement of a microlens is shown in Figure 2.36, including wavefront aberration, interference fringe pattern, and modulation transfer function (MTF).

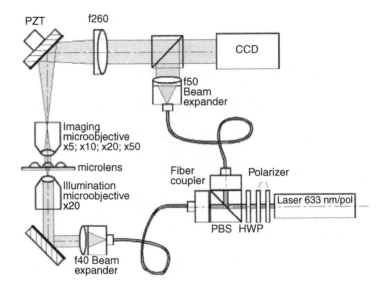

Figure 2.34   The schematic of a fiber-based Mach–Zehnder interferometer system.

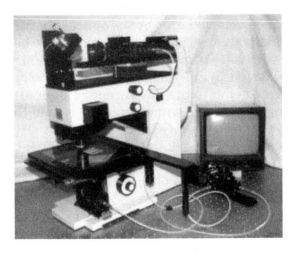

Figure 2.35   View of a fiber-based Mach–Zehnder interferometer
system.

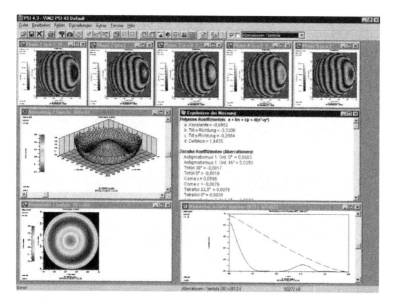

Figure 2.36 Example of wavefront aberration measurement in the Mach–Zehnder interferometer.

2.7.2.2.2 Shack–Hartmann Sensor

Figure 2.37 shows the principle of wavefront aberration measurement by application of the Shack–Hartmann sensor.[26] After optical radiation passed through a test microlens, it is divided from the different microlenses (this microlens is different from a test microlens, which is for Shack–Hartmann sensors) and

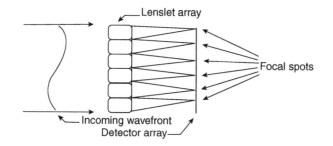

Figure 2.37 Principle of wavefront aberration detection in a Shack–Hartmann sensor.

is projected on sensor elements through a microlens, such as a CCD. The focus spot on the sensor element is focused from each microlens, which are shifted in the direction perpendicular to the incident wavefront. It is cleared for wavefront aberration by calculating the position of each divided spot.

This is a method for use as a wavefront-sensing system for adaptive optics to guarantee the disorder of telescopic performance measurement and the phase according to the fluctuation of the telescopic atmosphere. Further, use of a wavefront sensor with high perpendicular high detection sensitivity (the optical axis direction) is recommended for efficiency.

The procedure of wavefront measurement with a Shack–Hartmann sensor is simple. First, wavefront data are recorded as a reference without a test microlens in the optical pass. Next, a test microlens and a coupling lens are set into the optical pass, which uses a microscope objective lens, and wavefront data detection is performed.

However, the quality of the coupling lens must be checked in advance, which requires a good quality as a condensing lens. The aberration of the coupling lens can be disregarded when an error is small.

2.8 SUMMARY

Technology has improved a semiconductor process for microlenses, which are used for the newest technical applications. Microlens-related technologies were described, such as wafer-mounting technology and measurement technology for optical characteristics, including microlens surface structure measurement and wavefront testing of the microlens.

The next generation is progressing with the introduction of the microlens and its utilization. The simple form that includes an object to combine with a fiber (fiber array) and the microlens array for LCD projectors will continue as described here. A complicated form in a hybrid system will be recommended in the future.

Research and development aimed for application expansion will accelerate. The meaning, which also undertakes a realization of a fine and complicated form to the development of both wheels with evolution of measurement technology, is large.

REFERENCES

1. *Microlens Array Parts 1: Vocabulary*, ISO 14880-1, 1 (2001).

2. http://www.leister.com/microsystems/01a80d9291003e80c/ 01a80d9291005481e/index.html.

3. http://www.nsg.co.jp/en/device/index.html.

4. http://www.optical.ricoh.co.jp/precision/pre-03.html.

5. http://www.npl.co.uk/length/dmet/services/ms_microstruct.html.

6. http://www.leister.com/microsystems/01a80d9291003e80c/ 01a80d9291005481e/ index.html.

7. K. Goto, Y.-J. Kim, S. Mitsugi, K. Suzuki, K. Kurihara, and T. Horibe, Microoptical two-dimensional devices for the optical memory head of an ultrahigh data transfer rate and density system using a vertical cavity surface emitting laser (VCSEL) array, *Jpn. J. Appl. Phys.*, 41, 4835 (2002).

8. http://www.canyonmaterials.com/CMI-01-88-2.html.

9. http://www.canyonmaterials.com/request3.html.

10. S. Aoyama and T. Yamashita, Planar micro lens array using stumping replication method, *SPIE*, 3010, 277 (1997).

11. http://www.leister.com/microsystems/01a80d9291003e80c/ 01a80d93430807510/index.html.

12. http://www.leister.com/microsystems/01a80d9291003e80c/ 01a80d934214ca514/index.html.

13. http://www.nsg.co.jp/nsg/moc/pdf/pml.pdf.

14. http://www.memsoptical.com/prodserv/products/multi-func.htm.

15. Y. Sato, H. Mifune, Y. Kiyosawa, and S. Sato, Improvement of fabrication errors for a high NA microlens with two substrates, *Eighth Microoptics Conf. (MOC'01)*, H9, 162 (2001).

16. J. Schwider and O. Falkenstorfer, Twyman–Green interferometer for testing microspheres, *Opt. Eng.*, 34, 2972 (1995).

17. K. Miura and M. Okada, Three-dimensional measurement by a laser probe microscope with a scanning stage, *Proc. 28th*

Meeting Lightwave Sens. Technol., LST 28-1 (December 2001).

18. http://www.mitakakohki.co.jp/industry/nh_html/NH-3SP_Catalog.pdf.

19. D. Daly, *Microlens Arrays*, Taylor and Francis, London, pp. 101–102, 2001.

20. K. Hamanaka, H. Nemoto, M. Oikawa, and E. Okuda, Aberration properties of the planar microlens array and its applications to imaging optics, *Proc. SPIE*, 1014 Micro-Optics, 58 (1988).

21. T. Miyashita, K. Hamanaka, M. Kato, E. Sato, K. Nishizawa, and T. Morokuma, Wavefront Aberration Measuring Method for Microlens Using the Mach–Zehnder Interferometer Combining With an Effective Aperture (I), Institute of Physics, London, and Micro Optics and Lens Arrays, September 14, 2001.

22. http://physics.iop.org/IOP/Confs/MOLA/.

23. T. Miyashita, K. Hamanaka, M. Kato, S. Ishihara, H. Sato, E. Sato, T. Morokuma, Wavefront aberration measurement technology for microlens using the Mach–Zehnder interferometer provided with a projected aperture, *Proc. SPIE*, 5532, Interferometry XII: Applications, 117 (2004).

24. M. Kato, T. Miyashita, K. Hamanaka, S. Ishihara, E. Sato, T. Morokuma, Equivalence between the software-determined and the hardware-determined effective numerical aperture in the interferometrica measuring of microlens, *Proc. SPIE*, 5532 Interferometry XII: Applications, 136 (2004).

25. H. Sickinger, J. Schwider, and B. Manzke, Fiber based Mach–Zehnder interferometer for measuring wave aberrations of microlenses, *Optik*, 110, 239 (1999).

26. http://www.wavefrontsciences.com/semiconductor/columbus.html.

3

GRIN Lenses

KENJIRO HAMANAKA and HANK HASHIZUME

CONTENTS

3.1 INTRODUCTION

In classical optics, light rays travel in straight lines at a speed governed by the refractive index of the medium and in a direction dictated by Snell's law of refraction. In a medium in which the index is not uniform, however, light rays travel in a curved path from areas of lower index toward areas of higher index.

Gradient index (GRIN) lenses based on such a curved ray propagation have been widely used as collimator or coupling lenses in the field of optical interconnection. Selfoc®, trademarked GRIN lenses produced by Nippon Sheet Glass Company Ltd. (NSG), is a typical example of the GRIN rod lens fabricated using an ion exchange technique for glass material. The main reason for the Selfoc microlens (SML) to be adopted into various optical components in the telecommunications system is its unique shape; that is, in spite of a lens, the surfaces are flat, which provides compatibility with optical fibers, and the side is cylindrical, which enables good settlement with holders.

In addition, its excellent optical performance and its mass productivity may attract customers. Originally, the first multiglass composition GRIN fiber, which was the origin of Selfoc, was developed in 1968 using ion exchange technology in cooperative research between NSG and NEC Corporation.[1] Although manufacturing of GRIN fiber has taken the place of the vapor epitaxial method (e.g., vapor-phase axial deposition [VAD] or chemical vapor deposition [CVD]) nowadays, the fabrication technique that was applied to make the SML was considered to be a shorter shape of the GRIN fiber.[1-4] Planar microlens (PML) arrays derived from the technique are also considered promising for array applications in the field.

3.2 THE GRIN ROD LENS

3.2.1 Manufacturing Process

As rod glass is immersed in molten salt, the refractive index forms ion A+ in the glass, which comes out of the glass into the molten salt, as the ion B+ in the molten salt goes into the glass composition. Moreover, ion A+ in the glass moves internally from the center to the outside, induced by the ion

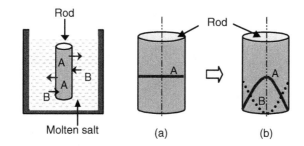

Figure 3.1 Schematic of the ion exchange method.

concentration difference generated by the first step. Finally, an ion concentration distribution in the radius is generated in the case of rod glass or it occurs along the depth in the case of plate glass, so that a refractive index distribution corresponding with ion distribution is generated in the glass (Figure 3.1). In fabricating GRIN lenses with the ion exchange method, ion migration occurs by the following two principles:

Ion exchange equilibrium at glass–molten salt interface

$$K = \frac{a'_B * a''_A}{a'_A * a''_B} = \frac{(G'_B * X'_B) * (G''_A * X''_A)}{(G'_A * X'_A) * (G''_B * X''_B)} \tag{3.1}$$

where a'_A, a'_B = ion activity in glass; a''_A, a''_B = ion activity in molten salt; G'_A, G'_B, G''_A, G''_B = activity coefficient; X'_A, X'_B, X''_A, X''_B = molar fraction; $X'_A + X'_B$ = constant; and $X''_A + X''_B$ = constant.

Ion mutual diffusion inside the glass

$$J_i = -\frac{D_A * D_B}{D_A * C_A + D_B * C_B} * \frac{d(\ln a_i)}{d(\ln C_i)} * \frac{dC_i}{dX} \tag{3.2}$$

where J_i = flux density of ion i (= $_{A, B}$); D_A, D_B = self-diffusion coefficient; C_i = concentration; a_i = ion activity; and dC_i/dX = concentration gradient.

In the case of the ideal situation ($K = 1$, $G'_i = 1$), the ion concentration in glass should be equal to that in molten salt, but in practice, ion selectivity (depending on the glass composition) relies on the interaction of other ions. Diffusion speed

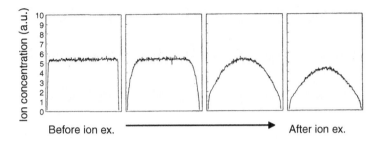

Before ion ex. ⟶ After ion ex.

Figure 3.2   Radial ion distribution in an SML.

and ion concentration distribution generated by ion diffusion internally are also affected by these interactions.

Returning to the SML, there are two types of refractive index distributions in a wide sense: RGI, which has a radial index change, and ZGI, which has an axial index change.[5] Here, we discuss RGI-type SML, which is widely used for fiber-optic components.

The fabrication process is as follows. First, the raw oxide glass rods are formed by drawing from the preformed bulk glass. Then, the glass rods are immersed in a molten salt bath, in which monovalent ions that give a higher refractive index in the glass composition are exchanged for ions that give a lower refractive index during this ion exchange process. Figure 3.2 shows the ion profile that induces a higher refractive index as measured by an x-ray microanalyzer (XMA) as ion exchange progresses. After the ion exchange process, a three-layer metal thin film that has a gold outer layer is coated on the side of the rod if it is needed for soldering with holders. Then, the glass rods are cut into appropriate lengths based on the lens pitch; both facets of the lens are polished to be optical-grade surfaces. Perpendicular flat, angled flat, or convex surfaces could be prepared. Finally, an antireflection coating is formed on each facet.

The refractive index is highest in the center of the lens and decreases with radial distance from the axis. The following equation describes the refractive index distribution of an SML.

$$N^2(r) = N_0^2 \{1 - (gr)^2 + h_4 (gr)^4 + h_6 (gr)^6 \ldots \} \qquad (3.3)$$

where r = distance from the optical axis, and N_0 = refractive index on the axis.

Equation 3.3 gives the high-order coefficients, such as h_4 or h_6. These numbers are important for the design of a lens that generates a diffraction limit optical spot.[2] For normal applications, such higher order coefficients can be negligible. Therefore, Equation 3.3 can be simplified to the following equation:

$$N(r) = N_0\left\{1 - \frac{(\sqrt{A})^2}{2}r^2\right\} \tag{3.4}$$

where $A = 2g$, and $\sqrt{A}$ = gradient constant.

In a GRIN lens, rays follow sinusoidal paths until they reach the back surface of the lens. A light ray that has traversed one pitch has traversed one cycle of the sinusoidal wave that characterizes the lens. Viewed in this way, the pitch is the spatial frequency of the ray trajectory.

$$2\pi P = \sqrt{A}Z \tag{3.5}$$

where P = pitch, and Z = the mechanical length of the lens.

Equation 3.5 relates the pitch to the mechanical length of the lens and the gradient constant. A larger $\sqrt{A}$ results in a pitch with a shorter lens length.

Using these parameters, ray trajectory in a GRIN lens can be simulated. All common paraxial distances can be derived from ray trace matrices. Assuming the optical system shown in Figure 3.3, the output ray height and slope are

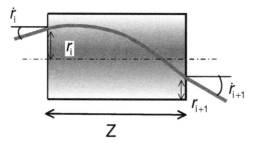

Figure 3.3 Ray trajectory.

described by Equation 3.6. In the case of a quarter pitch, the ray matrix is simplified as Equation 3.7.

$$\begin{bmatrix} r_{i+1} \\ \dot{r}_{i+1} \end{bmatrix} = -n \begin{bmatrix} r_{i+1} \\ \dot{r}_{i+1} \end{bmatrix} = \begin{bmatrix} \cos(\sqrt{A}Z) & \dfrac{1}{n_0\sqrt{A}}\sin(\sqrt{A}Z) \\ \cos(\sqrt{A}Z) & \dfrac{1}{n_0\sqrt{A}}\sin(\sqrt{A}Z) \\ -n_0\sqrt{A}\sin(\sqrt{A}Z) & \cos(\sqrt{A}Z) \\ \times \sqrt{A}\sin(\sqrt{A}Z) & \cos(\sqrt{A}Z) \end{bmatrix} \begin{bmatrix} r_i \\ \dot{r}_i \end{bmatrix}_0 \begin{bmatrix} r_i \\ \dot{r}_i \end{bmatrix}$$

(3.6)

$$\begin{bmatrix} r_{i+1} \\ \dot{r}_{i+1} \end{bmatrix} = \begin{bmatrix} 0 & \dfrac{1}{n_0\sqrt{A}} \\ -n_0\sqrt{A} & 0 \end{bmatrix} \begin{bmatrix} r_i \\ \dot{r}_i \end{bmatrix}$$

(3.7)

As mentioned, GRIN lenses are specified by the parameters N_0, $\sqrt{A}$, and P. To develop various types of SMLs, the glass compositions of the rod, combinations of ions in the molten salt, ion exchange conditions, and surface shape have been investigated (Table 3.1). For optical components in telecommunications systems, lens applications are roughly divided into two categories: optical collimating of optical fiber output light and optical coupling between laser diode–photodiode (LD/PD) and optical fiber. The latter requires a high numerical aperture (NA), which is achieved with a larger $\sqrt{A}$ and spherical surfaces.

3.2.2   Optical Collimator

In optical communications systems, the optical collimators are widely used for getting enlarged collimated beams out of the optical fibers or for focusing the enlarged beams into the optical fibers. Most in-line optical fiber modules consist of two face-to-face optical

TABLE 3.1 Optical Parameters of the SML

				Wavelength (nm)	1300		1550		
Type	Dia. (mm)	R (mm)	NA on axis	N_0	√A (per mm)	N_0	√A (per mm)	Z (mm)	Pitch (ex.)
SLS	2.0	—	0.37	1.552	0.238	1.550	0.237	6.63	0.25
SLW	1.0	—	0.46	1.592	0.597	1.590	0.596	2.64	0.25
	1.8	—	0.46	1.592	0.327	1.590	0.326	4.82	0.25
	2.0	—	0.46	1.592	0.295	1.590	0.294	5.32	0.25
	3.0	—	0.46	1.617	0.199	1.615	0.199	7.89	0.25
	4.0	—	0.46	1.617	0.148	1.615	0.148	10.63	0.25
SLC	1.8	—	0.46	1.592	0.323	1.590	0.322	4.88	0.25
SLH	1.8	—	0.60	1.636	0.418	1.634	0.417	3.77	0.25
PCW	1.8	2.0	0.60	1.592	0.327	1.590	0.326	4.81	0.25
PCH	1.8	2.0	0.61	1.636	0.418	1.634	0.417	3.01	0.20
PCT	1.8	1–3	0.57	1.636	0.392	1.634	0.391	3.26	—
PCK	1.8	1–3	0.54	1.636	0.392	1.634	0.391	—	—

Note: Type, product code; Dia., diameter; ex., example.

collimators and of functional elements, such as filters, polarizers, or crystals, inserted in the free space between them. Figure 3.4 is a schematic of an in-line optical isolator. By sandwiching a narrow-band optical thin-film filter (TFF) with a single-fiber collimator and a dual-fiber collimator, in-line optical filter modules can be fabricated. Then modules with a bit different center path wavelength are cascaded to assemble optical multiplexers or demultiplexers (Figure 3.5). For instance, 64 GRIN lenses could be used in the 32-channel component. These types of TFF multiplexers/demultiplexers are mainly used for dense wavelength division

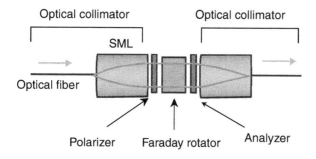

Figure 3.4 In-line optical isolator.

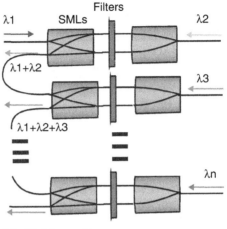

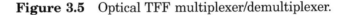

λ1+λ2+λ3+... +λn

Figure 3.5   Optical TFF multiplexer/demultiplexer.

multiplexing (DWDM) systems in metropolitan or long-haul applications with fewer than 40 channels; arrayed waveguide
(AWG)–type components tend to be used for long-haul or ultral-
ong-haul applications with more than 40 channels.

Figure 3.6 shows a structure of the actual optical fiber
collimator. One or two optical fibers are fixed into a glass
capillary with epoxy adhesive, and its end facet is polished
with 6 to 12° angles. The small gap between optical fiber and
lens remains, with no epoxy in the optical path after all the
parts are fixed with epoxy.

Using lenses with different focal lengths, various loss
characteristics can be achieved. There are convenient equations
to calculate the beam waist or the beam propagating length

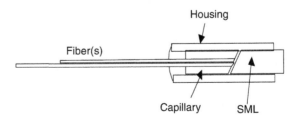

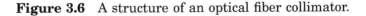

Figure 3.6   A structure of an optical fiber collimator.

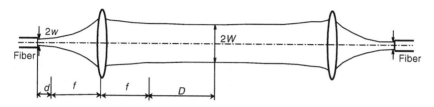

Figure 3.7 An optical coupling system with fiber collimators.

for low-loss coupling of optical fiber collimators. Equations 3.8 and 3.9 describe the Gaussian beam conversion well by assuming the optical coupling system shown in Figure 3.7.

$$D = f^2 d / \left(d_m^2 + d^2 \right) \qquad (3.8)$$

where $d_m = \pi w^2/\lambda$.

$$W = wf / \left(d_m^2 + d^2 \right)^{1/2} = w\,(D/d)^{1/2} \qquad (3.9)$$

where d and D = distances from the focal points of the lenses to beam waists, respectively; f = focal length; and λ = wavelength. To realize low-loss coupling, the beam waist positions as well as the beam spot diameters of the collimator pair should correspond. If coupling the same two collimators, the beam waist should be placed in the center of the working distance (WD), that is, the propagating free space. Equation 3.8 and Equation 3.9 introduce Figure 3.8.

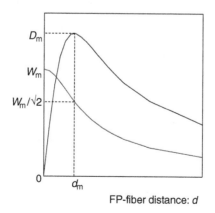

FP-fiber distance: *d*

Figure 3.8 A beam waist and its position depending on fiber position.

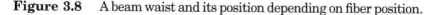

In the case of d = 0, 2W becomes maximum; then,

$$2W_m = 2(\lambda/\pi)(f/w) \qquad (3.10)$$

Therefore, when optical fibers with different mode field diameters 2w and 2w' as input/output ports, respectively, W and W' should be matched by selecting f' proportional with w'.

On the other hand, two d exist for one D. If d = d_m, 2D is maximum.

$$2D_m = (\lambda/\pi)(f/w)^2 = \pi(2W_m)^2/(4\lambda) \qquad (3.11)$$

For instance, it is obvious from Equation 3.11 that a beam waist diameter of at least 0.37 mm is needed to achieve low-loss coupling through a 70-mm free space distance at a wavelength of 1550-nm.

Figure 3.9 shows experimental results of such insertion loss characteristics depending on working distances, that is, lens-to-lens distance in two facing collimators for which the fiber–lens positions, including the distance from fiber to lens, were realigned for each working distance. Each lens has a 0.23 pitch. A loss characteristic of an actual C180 single-fiber

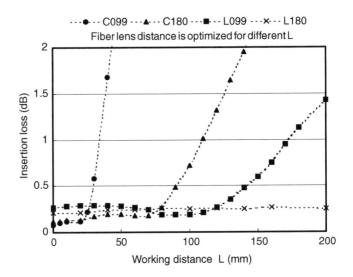

Figure 3.9 Insertion loss characteristics vs. WD.

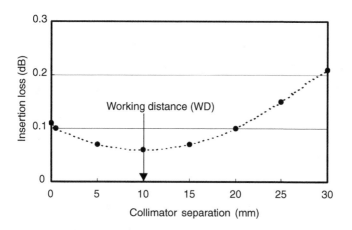

Figure 3.10 Loss characteristics of FCS-C180-23-010 (single-fiber collimator).

collimator optimized for a 10-mm WD (FCS-C180-23-010) is shown in Figure 3.10. In addition, Figure 3.11a and Figure 3.11b represent loss characteristics of a dual-fiber collimator optimized for 10-mm WD (FCD-C180-23-010). Insertion losses at the desired working distance of 10 mm are less than 0.15 dB for both single- and dual-fiber collimators, and a 0.02-dB loss tolerance can be achieved at ±5 mm.

Polarization-dependent loss (PDL), wavelength-dependent loss (WDL), and temperature-dependent loss (TDL) should

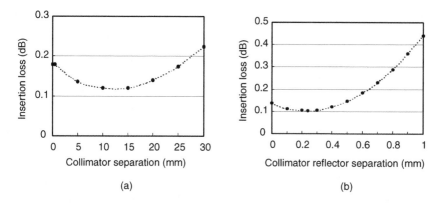

Figure 3.11 Loss characteristics of FCD-C180-23-010 (dual-fiber collimator). (a) Passing port; (b) reflection port.

also be important parameters for designing optical components using optical collimators. Such loss characteristics on
C180 collimators are reported in Reference 6.

The return loss caused by a back reflection from each part's
facet was less than –66 dB in the temperature range –20 to
+85°C.

3.2.3   Laser-to-Fiber Coupling

When the laser output light from an LD chip is coupled into
single-mode fiber (SMF), two types of lens system are mainly
applied. One uses a single PC (plano-convex)–SML, and the
other is a combination of a PC–SML and a ball lens, aspherical
lens, or PP (plano-plano)–SML. The latter is for placing some
functional elements, such as an isolator, in the optical path.
The light emitted from an LD has three times the radiation
angle than that from an SMF. Therefore, the first lens placed

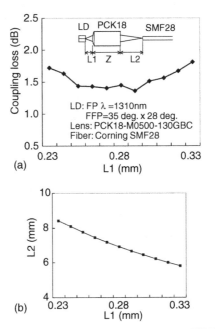

Figure 3.12   Fabry–Perot laser diode (FP–LD)/SMF coupling characteristics using a PCK lens.

close to an LD chip should have a larger NA, larger effective area, and well-controlled higher order coefficients of its refractive index distribution to get higher coupling efficiency. These lenses are premised on use with a spherical surface that generates most of the lens power, and its spherical aberration can be compensated with the gradient refractive index in the lens. It is achieved by controlling the wavefront form of the transparent light through the lens.

Figure 3.12 shows optical coupling characteristics between LD and SMF using a Selfoc lens. It is a Fabry–Perot (FP) laser with a 1310-nm center wavelength and 35° × 28° radiation angles. Corning SMF28 fiber was used for this measurement. The coupling loss was good as low as 1.4 dB even though it was a single-lens system.

3.3 PLANAR MICROLENS ARRAYS

There has been a great deal of attention paid to arrayed fiber-optic devices because of the rapid development of next-generation multichannel optical communications systems. A PML array is considered a key component in such an arrayed application. The PMLs are fabricated by applying the ion exchange process of the SMLs into the planar substrate instead of the glass rod.[7,8] Figure 3.13 is a photograph of a

Figure 3.13   A photograph of a PML.

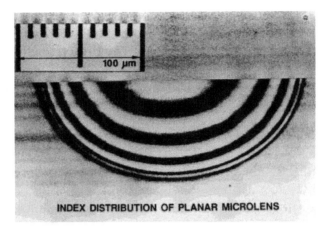

INDEX DISTRIBUTION OF PLANAR MICROLENS

Figure 3.14   A refractive index profile of a PML.

PML that is a GRIN microlens array formed on a glass substrate. Each microlens has a hemispherelike index profile (Figure 3.14). Precisely aligned flat surface microlens arrays are realized by combining the ion exchange process with a photolithographic technique. Specific glass material with a 1.523 refractive index at a wavelength of 1.55 μm is used as the substrate, and the central part of the lens region results in n = ~1.661.

   Figure 3.15 is a schematic diagram of the PML fabrication process (see the right part corresponding to "flat type"). First, a metal mask is coated and patterned by the photolithographic technique; then, the ion exchange is carried out. Higher refractive index ions are diffused into the glass substrate via the circular windows of the metal mask. As a result, a hemispherelike profile of higher refractive index is formed that works as each microlens. Fabrication conditions such as composition of the molten salt, temperature of the salt, and diameter of the windows are chosen to optimize the refractive index profile so that the wavefront aberration becomes small enough to correspond to the specifications of the lenses to be fabricated. After the ion exchange process, the metal mask is removed by etching; then, the glass surface is polished to flatten the slightly swelled part inside the windows that occurred during the ion exchange process.

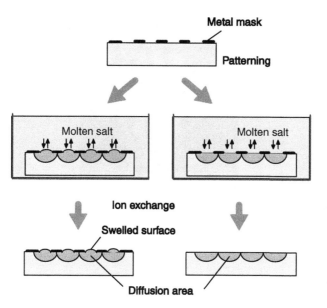

Figure 3.15 Fabrication process for PMLs.

For fabricating the specific PMLs for the individual customer, the ion exchange time and the window diameter are designed for a specific lens diameter. The composition of the molten salt as well as the polishing condition are chosen to be suitable for the specific NA. It is obvious that the photomask design determines the specific alignment of the lens array. It is also available for fabricating an angled facet at the back surface with an accuracy of ±0.5°, for instance.

Figure 3.16 is an interferogram of a PML with a 0.25-mm diameter and 0.65-mm focal length. The wavefront aberration is measured as ~0.04 in root mean square (RMS) value. Figure 3.17 shows the accuracy of center-to-center spacing of the PMLs. The accuracy is within ±0.4 μm for each lens. The off-axis aberration property of the PML is also reported briefly in Ref. 9.

Figure 3.18 shows a schematic figure of an example for applying the PMLs. The PMLs are considered promising as collimation lens arrays in optical-switching modules. They can also be used as NA conversion lens arrays in hybrid optical

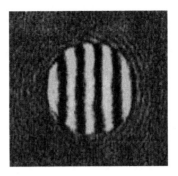

Figure 3.16   An interferogram of a PML with a 0.25-mm diameter.

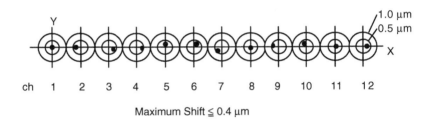

Figure 3.17   The accuracy of center-to-center spacing of the lenses.

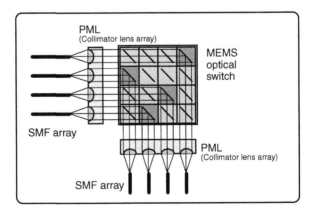

Figure 3.18   An application of the PML for a collimator lens array.

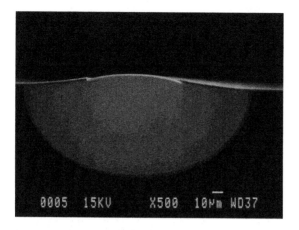

Figure 3.19 A high NA-type PML combining gradient index with swelled surface.

systems that can be used in multiplexer/demultiplexer modules, for instance. PMLs with a flat surface and an NA about 0.15 to 0.19 are suitable for applications combining SMFs.

On the other hand, lenses with higher NAs are required for coupling applications between LD arrays and fiber arrays. For this purpose, we have also developed swelled-type PMLs (SPMLs) that combine a spherically swelled surface with the refractive index profile.[10] Because of the difference of the ion diameters to be exchanged, the surface of the window region is swelled during the ion exchange process. The SPMLs control the swelled profile carefully so that a higher NA with low aberration is achieved (see the left part of Figure 3.15, corresponding to the SPML). Figure 3.19 shows a photograph of the SPML with an 85-µm diameter and 115-µm focal length and an NA of 0.37. It can be used in coupling applications of LDs and SMFs (Figure 3.20). In addition to reasonable coupling efficiency, alignment tolerance can be widened while using the SPML as the coupling lenses. Several durability tests corresponding to Telcordia standards, for instance, have been carried out for both the flat-type PMLs and the SPMLs; all tests were successfully finished. Several PMLs with certain specifications, including both flat-type and SPMLs, have been under mass production.

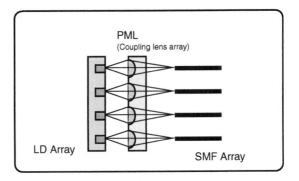

Figure 3.20   An application of the SPML.

We have also fabricated a fiber collimator module combining eight-channel SMFs with eight-channel PMLs with a
0.7-mm focal length at a 1550-nm wavelength.[11] A fiber array
(eight SMF28 fibers in a Pyrex V-groove array) was aligned
to the PML and then fixed together using epoxy adhesive.
The dimensions of the resultant fiber collimator array were
$11 \times 5 \times 4$ mm ($L \times W \times H$). Figure 3.21 shows the collimation
property of the fiber collimator module, by means of insertion
loss vs. collimation length. Low insertion loss of around
0.5 dB was obtained at a collimation length less than 10 mm,
which is equal to the theoretical limitation of Gaussian beam

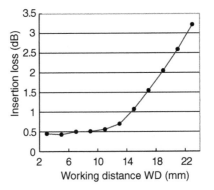

Figure 3.21   The collimation property of an eight-channel fiber
collimator array module using the PML.

propagation using the f = 0.7 mm lenses used in the experiment. The $1/e^2$ beam diameter is about 0.1 mm. Interchannel variation was ~0.12 dB. The crosstalk from the other channels was less than −45 dB, and return loss was above 60 dB.

Demands of the microlens array have increased in both DWDM and local-area network (LAN) applications. Precisely aligned microlens arrays with high optical performance as well as high durability are promising. PMLs fabricated by the ion exchange technique are the most attractive candidate for such microlens arrays.

3.4 CONCLUSIONS

GRIN lenses manufactured by ion exchange methods have been a viable solution for diode-to-fiber coupling and are the preferred solution for fiber-to-fiber coupling in the telecommunications as well as the optical interconnection field for many years. Advances in the ion exchange method and the integration of new technologies will ensure that the GRIN technology continues to provide solutions to the fields.

REFERENCES

1. T. Uchida, M. Furukawa, I. Kitano, K. Koizumi, and H. Matsumura, *IEEE J. Quantum Electron.*, QE-6, 606 (1970).

2. I. Kitano, H. Ueno, and M. Toyama, Gradient-index lens for low-loss coupling of a laser diode to single-mode fiber, *Appl. Opt.*, 25, 3336–3339 (1986).

3. T. Nishi, H. Ichikawa, M. Toyama, and I. Kitano, *Appl. Opt.*, 25, 3340–3344 (1986).

4. I. Kitano, H. Ueno, and M. Toyama, *Appl. Opt.*, 25, 3336–3339 (1986).

5. P. McLaughlin, M. Toyama, and I. Kitano, *SPIE*, 695, 194–198 (1986).

6. H. Hashizume, K. Hamanaka, A.C. Graham, and X.F. Zhu, The Future of Gradient-Index Optics, Proceedings of the SPIE Annual Meeting, July 31–August 3, 2001, paper 4437-05.

7. K. Iga, M. Oikawa, S. Misawa, J. Banno, and Y. Kokubun, Stacked planar optics: an application of the planar micro-lens, *Appl. Opt.*, 21, 3456–3460 (1982).

8. M. Oikawa, K. Iga, T. Sanada, N. Yamamoto, and K. Nishizawa, Array of distributed-index planar microlens prepared from ion exchange technique, *Jpn. J. Appl. Phys.*, 20, L294–L298 (1981).

9. K. Hamanaka, H. Nemoto, M. Oikawa, E. Okuda, and T. Kishimoto, Multiple imaging and multiple Fourier transformation using planar microlens arrays, *Appl. Opt.*, 29, 4064–4070 (1990).

10. M. Oikawa, H. Nemoto, K. Hamanaka, and T. Kishimoto, Light coupling between LD and optical fiber using high NA planar microlens, *Proc. SPIE*, 1219, 532–537 (1990).

11. X. Zhu, Y. Sato, Y. Sasaki, and K. Hamanaka, Fiber collimator arrays for optical devices/systems, *Proc. SPIE*, 4564–4519, Opto-Mechatronic Systems (2001).

4

Development of Diffractive Optics and Future Challenges

KASHIKO KODATE

CONTENTS

4.1 INTRODUCTION

The advent of the information era in the 21st century promises a prosperous future and improved welfare for all. Among others, optical technology is expected to play a key role, and its importance is increasing rapidly. Conventional optical technology has centered around optical information and communication, with a major focus on the development of optical methods for electronic information processing as well as the fusion of technologies. Examples include the development of interfaces between optical and electronic technology and elements or components for controlling wave planes. There is also an increasing demand for optical technology in a wider range of fields, such as medicine, social welfare, and the environment.

Accordingly, the transition from electronics to optics is becoming common and desirable. All-optical systems offer new functions, such as multiplication and parallelization, with less restriction between devices because of the elimination of optoelectronic conversion.

In general, the control of information using optics requires optical devices that provide the functions of wave plane control, amplitude control, polarization of light, and wavelength conversion (Table 4.1). First-generation bulk elements such as lenses, mirrors, and prisms have been used extensively to achieve these functions, and the fabrication methods for these elements have been well established. Nonetheless, it remains difficult to miniaturize the optical apparatus further and to achieve multiple functions because of the limitations of precision for conventional fabrication techniques. The second generation of optical elements has sought to replace bulk elements with compact, lightweight microoptical elements.

Compared to refractive microoptical elements (e.g., microlens, rod lens), diffractive optical elements (DOEs) have attracted attention as devices of high utility because of their

TABLE 4.1 Evolution of Optical Elements

	First generation	Second generation	Third generation
Theory	Conventional optics	Microoptics	Integrated optics
Light source	Gas lasers	Semiconductor lasers	Semiconductor lasers
Optical element	Lenses, mirrors, prisms	Lens arrays, zone plate	Waveguide devices for optical interconnection
Optical system	Free-space optics, mechanical controlling	Integrations, miniaturized mechanics, micromechanics	Integrated waveguide optics, fiber-optic lenses, optical interconnections
Examples	Holographic recording	4 × 4 Rectangular lens array	Planar interconnection module

relatively low spherical aberration and increased latitude in design, allowing integration of an entire system, including LD light sources, optical fiber waveguides, and photodetectors. DOEs represent a key technology in third-generation integrated optics.[1]

DOEs are composed of uneven surfaces and have a refractive index and amplitude distribution according to the material used, allowing effective control of wave planes. Although high diffractive efficiency cannot be obtained with two-dimensional (2D) planes, the fabrication of three-dimensional (3D) configurations utilizing the Bragg condition can improve the utility of DOEs by acting as wavelength selectors. The 3D configurations can be achieved by controlling the depth of the surface feature, and the Bragg condition can be set by controlling the blazed angle and grating depth.[2–4]

DOEs can be fabricated with or without periodicity. Computer-generated holograms (CGHs) are representative of the latter. Nanoscale DOE structures with periodicity smaller than the operating wavelength or the proximate sphere are expected to provide further improvement in the functionality and performance of optical systems. Such subwavelength gratings[5] are essentially the same as other types of gratings because the design and fabrication methods are common to all. However, this does not account for the quantum effects arising from the use of such small cycles. Although there is as yet no clear definition of diffractive optics, Figure 4.1 illustrates that the concept bridges a number of technologies.[6]

Although research on gratings and Fresnel zone plates (FZPs) in application to diffractive optics is not new, the advent of optoelectronics systems with the development of laser diodes (LDs) as light sources has gradually brought DOEs and their applications to the attention of researchers.[7,8] My group is considered one of the leaders in this field, having undertaken specific research in this area from an early stage.[9–11]

Binary optics,[12] a novel design approach in which the blazed configuration is approximated as a digital blaze, has attracted sudden attention from a wide range of research groups around the world. Binary optics represents a computer-based design approach, and fabrication is performed by such semiconductor

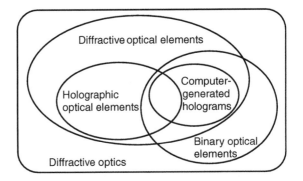

Figure 4.1   Concept of diffractive optics.

processes as lithography and dry etching, both of which are used widely as mass production techniques. This promotes efficiency by facilitating the production of surface relief DOEs with high diffraction rates.

More than 200 papers are presented annually at international conferences held by the Optical Society of America (OSA).[13] Although the OSA is primarily concerned with basic theory, design, and fabrication, other conferences that are more application related, such as those concerning information optics, photonic networks, and nanotechnology, are accepting an increasing number of articles on DOEs. In Japan, in addition to the Micro Optics Conference (MOC) held every 3 years,[14] the KOGAKU (in Japanese) Symposium of the annual conference of the Optical Society of Japan (OSJ) and a symposium organized by the Optics Design Group of OSJ represent the main arenas for researchers to contribute to progress in this field. At recent conferences, the focus was on new functions to improve the performance of DOEs, highlighting applications using various materials or new elements integrated with the refractive element.[15]

This chapter recounts the developments in DOEs and related spectroscopy, presenting the types and functions of DOEs, and it outlines the fundamentals of diffraction theory and numerical analysis. Fabrication techniques for DOEs are also explained, referring to research in pursuit of optimal design for higher efficiency of multilevel optical elements. In the section

on future challenges, a variety of recent applications are introduced. The representative examples described include an all-
optical switching module for high-speed signal processing,[16]
high-dispersion volume-phase holographic (VPH) grisms for
spectroscopic observation for the Subaru Telescope,[17] and a
facial recognition system.[18] Other prospects for optical information technology techniques are also presented.

To conclude, some of the obstacles and limitations of
these current applications are discussed as a guide for future
development in this field.

4.2 VARIETIES OF DIFFRACTIVE OPTICAL ELEMENTS

4.2.1 Classification of DOEs

Figure 4.2 exhibits classifications of two types of DOEs with
a periodic structure: amplitude type and phase type, which is
controlled by a phase modulator. Phase-type DOEs feature
higher diffraction efficiency; the relief type exceeds the refractive rate distribution type in its wider reception angle, minute
uneven surfaces of wavelength order, and availability.

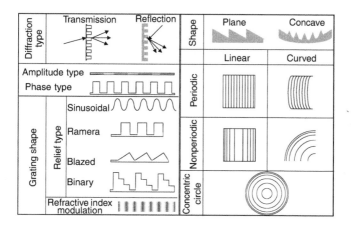

Figure 4.2 Classification of diffractive optical elements.

4.2.1.1 Diffractive Gratings

The diffraction grating[19] is the most established diffracting optical component. It may be regarded as the diffraction equivalent of the prism. The gratings and the prisms serve the same function, but in many important regions of the spectrum, the grating performs better. It is for this reason that so much effort has been devoted to ruling gratings by mechanical means, a task that demands the highest possible precision and skill. In 1786, the dispersion of light into spectral colors by a periodic structure was first reported by Rittenhouse, an American astronomer.[20] But it fell in 1812 to Fraunhofer to reintroduce the idea of grating.[21] Since Fraunhofer, of course there have been many developments in the theory and manufacture of diffracting gratings and their utilization in spectroscopic and optical instruments. Remarkable among the many contributors are Rowland, who invented the concave grating and constructed precision ruling engines,[22] and Wood, the first to produce gratings with a controlled groove shape, the "blazed" gratings.[23]

Among other periodic structured elements, diffraction gratings are well known. They apply diffraction and interference of the light wave that permeates or reflects through slits (Figure 4.3). The fundamental feature of a diffraction grating is its periodic structure: The transmission function is a periodic function of position.

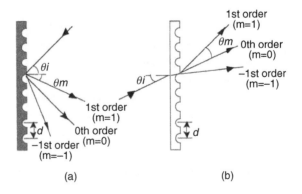

(a)   (b)

Figure 4.3 Principle of two types of diffractive gratings: (a) reflection and (b) transmission phase.

Considering a one-dimensional (1D) case for simplicity, if the transmission function is denoted by $t(x)$, where x is the position on the grating, and the grating period by d, the periodicity of the grating means $t(x) = t(x + d)$. A conventional method of analysis for periodic functions is the use of Fourier series. For the function $t(x)$ of period d, the Fourier series representation of $t(x)$ is

$$t(x) = \sum_{m=-\infty}^{\infty} C_m \exp(i2\pi m f_0 x) \tag{4.1}$$

where $f_0 = 1/d$ is the grating spatial frequency, and the coefficient C_m is given by

$$C_m = \frac{1}{\Lambda} \int_0^{\Lambda} t(x) \exp(-i2\pi m f_0 t \dot{x}) dx \tag{4.2}$$

If $t(x)$ is a purely real function, the grating is referred to as an amplitude grating; if $t(x)$ is a unit-modulus complex function, the grating is a phase grating.

In general, $t(x)$ can have both amplitude and phase components, but because the grating is a passive element, $|t(x)| \leq 1.0$.

It is evident from the grating Equation 4.3. that the directions of a diffracted order depend on the wavelength. Consider the somewhat general situation of oblique incidence (Figure 4.4). The grating equation for both transmission and reflection becomes

$$d(\sin\theta_i \pm \sin\theta_m) = m\lambda \quad (\text{with } m = 0, \pm1, \ldots) \tag{4.3}$$

where θ_i denotes incident angle, θ_m is mth order diffraction angle, λ is the wavelength of the input wave, and d is the period of diffractive grating. This expression applies equally well regardless of the refractive index of the transmission grating itself.

Concerning performance of gratings with a periodic structure, diffraction efficiency has significance. Diffraction efficiency

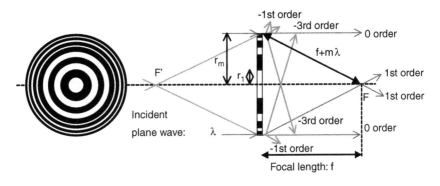

Figure 4.4 Configuration of Fresnel lenses. Full-period Fresnel zone construction on a diffractive lens. The path length to the focal point incrementally increases from zone-to-zone by one wavelength.

is shown in the light intensity ratio measured against the incidence wave centered on the diffractive wave.

$$\eta_{\pm m}(\%) = (I_{\pm m}/I_{\text{in}}) \times 100 \qquad (4.4)$$

where $I_{\pm m}$ is the mth order diffractive wave intensity, and I_{in} is the incidence wave intensity.

If we consider the sample example of the Littrow mounting in which light is diffracted back close to the path of the incident beam $\theta_{\text{in}} = \theta_m = \theta$, the grating equation becomes

$$2d \sin \theta = m\lambda \qquad (4.5)$$

from which it follows that

$$\frac{\partial \theta}{\partial \lambda} = \frac{m}{2d \cos \theta} \qquad (4.6)$$

The angle by which we must rotate the grating to scan a given element of the spectrum is therefore just half of the angle subtended by the same spectrum in a spectrograph.

4.2.1.2 Diffractive Lenses

The first optical element to use diffraction to focus light and form images like a lens was the device known today as a diffractive lens.[24] If the structures within each of the annular zones consist of a continuous phase function, the lenses are generally called *Fresnel lenses*. In FZPs, the ring system consists of diffracting binary phase or amplitude structures. The zone plate consists of alternating transparent and opaque annuli, with radii selected such that the distance from the edge of each annulus to the focal point is an integral number of half-wavelengths longer than the axial focal length shown in Figure 4.4.

The incidence of the plane wave with the wavelength perpendicular to the device promotes the formation of images at the focal distances F and F' by the light from the respective transparent part because of its diffractive effects by a circular zone and provides it with the lens function. The focal length of the wavelength λ is $f = \pm r_m^2 / m\lambda$; high-order diffractive light creates subfocal points at the distance denoted by $\pm f/(2m + 1)$ (where m is an integer) in addition to the main focal points. However, the intensity of image formation decreases in proportion to $1/(2m + 1)^2$ as the value of m increases.

Because the optical path length between the mth wave from the circular zones and the light crossing the light axis is the multiple of the wavelength by any integer numbers, the two lights reinforce each other. The radius of the circle for the mth r_m is given by

$$\sqrt{r_m^2 + f^2} - f = m\lambda \tag{4.7}$$

where f is sufficiently larger than r_m, Equation 4.7 can be denoted as

$$r_m^2 = mf\lambda \quad (m = 1, 2, \ldots, N) \tag{4.8}$$

Therefore, the width of the mth circle W_m is provided by

$$W_m = \sqrt{r_m} - \sqrt{r_{m-1}} \tag{4.9}$$

The Fresnel lens in Figure 4.5b is an optimized blaze type, with the same phase information removed from a conventional refractive lens, through which nearly 100% efficiency can be attained.

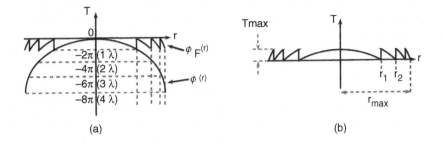

(a)   (b)

Figure 4.5   Transfer from a refractive lens to a diffractive lens;
(a) refractive phase function; (b) devised diffractive lens.

A Fresnel lens with an incidence light wavelength and
a focal length has a phase shift function $\phi(r)$, indicated by
Equation 4.10, where f is considerably larger than a lens aperture $(f > r)$. The equation is provided as follows:

$$\phi(r) = \frac{2\pi}{\lambda}\left(f - \sqrt{f^2 + r^2}\right) \tag{4.10}$$

$$\phi(r) = -\pi r^2 / \lambda f \tag{4.11}$$

It is worth noting, however, that the feasible volume level of
the phase shift remains within the realm of 0 to 2π, which
leads to another equation of phase shift function:

$$\phi_F(r) = \phi(r) \bmod 2\pi \tag{4.12}$$

To realize this phase shift, a relief-type DOE is desirable
owing to its facility in adjusting thickness.

The depth of relief is

$$D(r) = \phi_F(r)\lambda / 2\pi(n - 1) \tag{4.13}$$

where n is the refractive index.

The radius and width of zones is equal to those of FZPs.

The brightness of the lens is denoted by its numerical
aperture (NA):

$$NA = r/f \tag{4.14}$$

From Equation 4.14, it is shown that the lens becomes
brighter with increasing NA. The image formation spot can

be obtained by the following equation derived from the basic one for the Gaussian beam:

$$\omega_{min} = \lambda/L \cdot NA \qquad (4.15)$$

Performance of a diffractive lens is designated by its diffraction efficiency, which is endowed with first-order diffractive light in Equation 4.4.

4.2.1.3 Digital Blazed Optical Elements

In 1985, a DOE was developed at Massachusetts Institute of Technology's (MIT's) Lincoln Laboratory. With computer-generated design data and integrated circuit (IC) fabrication techniques, robust and efficient DOEs could be built. This breakthrough was first presented at SPIE in 1988 in a paper, "Diffractive Optical Elements for Use in Infrared Systems." The experiment by Swanson and Veldkamp proved that digitally blazed planes on a convex lens could reduce the color and spherical aberration of the infrared silicon single lens. The term *binary optics* was coined on that occasion.[12,25]

As in Figure 4.6, the multilevel zone plate (MLZP)[26] is a device of which a sawtooth-shaped cross section of the Fresnel lens is digitally approximated by increasing phase levels formed in M process steps from $M + 1$ to 2^N, greatly reducing the number of process iterations and processing costs needed to fabricate DOEs with high diffraction efficiency. When the level number is M, the depth of a zone is denoted by the following

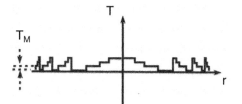

Figure 4.6 Illustration of a multilevel approximation to the diffractive surface.

formula, given that the volume of phase shift T_{max} of 2π is equally divided into 2^M steps.

$$T_M = 2\pi\lambda/(n-1)2^M \qquad (4.16)$$

A radius of the mth zone can be deduced from the equation for the radius of MLZP.

$$r_m = \sqrt{\frac{2mf\lambda}{M} + \left(\frac{m\lambda}{M}\right)^2} \qquad (4.17)$$

An image formation device for optical communications requires high NA, efficiency, and resolution. Thus, the level number N or radius r has to be increased, and the line width of the most exterior has to be minimized according to the increased level. The limit of line width W_{min} is dependent on the quality of the fabrication apparatus.

4.2.1.4 Holographic Optical Elements

Holography was invented in 1948 by Gabor,[27] who was looking to improve the quality of electron microscopic images. He demonstrated that it was possible to record and reconstruct a complex optical wave front with full information (i.e., the amplitude as well as the phase information) by recording the interference pattern of the wave front with a reference wave front. He called this process *holography* (from the Greek *holos* meaning "complete," referring to the recording of the entire complex wavefront).[28]

In 1962, Leith and Upatnieks proposed a fabrication process after successfully recording two bundles of laser lights in parallel as interference patterns on Kodak 649F.[29] Incoherence of thickness of emulsion and distortion of the hologram plate incurred by developing and drying generated aberration, resulting in poor performance as a grating. Subsequently, the quality was refined by development and use of high-resolution and nongranular photosensitive polymer and photoresist.

A spectrometer with a holographic diffractive grating is now commercially provided by Jobin-Yvon Corp.[30] Nonspherical exposure and control of the grating structure and integrating

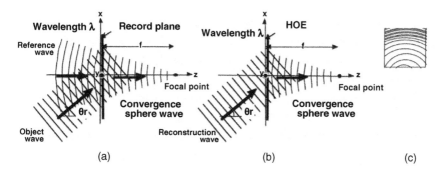

Figure 4.7 Fabrication process for HOEs. (a) fabrication by two-beam interferences; (b) reconstruction of HOE; (c) holographic pattern.

distance became feasible, which was not possible by an engine with conventional ruling. This development contributes to aberration corrections and removal of stray light, leading to better spectrometer performance. Moreover, products with higher diffraction efficiency are fabricated by blazing a cross section of a sinusoidal formation with the ion beam etching method.[31]

As a basic example of holographic optical elements (HOEs), an off-axis type is shown in Figure 4.7. The HOE shows interference of the plane wave at the angle θ_r toward the record plane and spherical wave converging at distance f behind the record plane. This hologram performs a function of a lens with a focal length f, which collects plane waves with identical wavelength and incidence angle as the ones at the recording moment.

More frequently, holographic recording techniques are used for the fabrication of volume grating.[32] For this purpose, thick ($t > 10$ μm) layers of photoemulsion are used to record the interference fringes. The thick HOEs diffract light with high efficiency into a single diffraction order, which is their most important property. It is also possible to achieve large diffraction angles because of the holographic recording process. These are obviously attractive features for applications in microoptics. As described, thick HOEs are optimized to yield nearly 100% efficiency for reconstruction under the Bragg condition.

4.2.1.5 Numerical-Type DOEs

A numerical-type DOE (CGH)[33] is calculated and optimized as a 2D matrix of regularly spaced complex data sampled over x–y space. This sampled representation is used to optimize a DOE when none of the previous analytical methods can provide the design engineer with an adequate solution to problems. Usually, the optical functions performed by numerical-type elements are more complex than the optical functions performed by analytical-type elements (DOEs). Among the typical functions that can be implemented by numerical-type DOEs are beam shaping, Fourier filtering, 2D and 3D display, and spot array generators.

If the z-axis is designed to be perpendicular to the hologram plane (x,y), the wave plane of the convergent spherical wave at the recording time can be denoted as follows:

$$\phi_s(x,y) = -k\sqrt{x^2 + y^2 + f^2} \qquad (4.18)$$

The sphere representing the wave plane, which proceeds at the angle θ_r on the hologram plane, is

$$\phi_r(x,y) = -k\sqrt{x^2 + y^2 + f^2} \qquad (4.19)$$

Each line of the interference patterns can be described as

$$\phi_r(x,y) = -k\sqrt{x^2 + y^2 + f^2} \qquad (4.20)$$

where m is the line number of the interference patterns, and c is constant.

$$\sqrt{x^2 + y^2 + f^2} + x\sin\theta_r = m\lambda + f \qquad (4.21)$$

$$(m = ..., -2, -1, 0, +1, +2, ...)$$

Assuming that Equations 4.18 and 4.19 are now substituted for Equation 4.20 and c is chosen so that the line ($m = 0$) should pass the origin, Equation 4.22 is acquired.

Equation 4.22 represents the ellipse group, and the answer x is

$$x = \frac{\sqrt{f^2 \sin^2 \theta_r + m^2 \lambda^2 + 2m\lambda f - \gamma^2 \cos^2 \theta_r} - (f + m\lambda)\sin\theta_r}{\cos^2 \theta_r}$$

(4.22)

From these equations, a fringe pattern will emerge (Figure 4.7c). The HOE is applied to such products as an image scanner.[34] Table 4.2 summarizes the advantages and limitations of numerical-type DOE (CGH) techniques.

4.3 FUNCTIONS OF DIFFRACTIVE OPTICAL ELEMENTS

DOEs possess a variety function of exchanging wave planes (Figure 4.8). There are four function classifications[35]:

1. Image formation, chiefly entailing image formation, concentration, collimate emission.
2. Diverging and converging wave multiplexing/demultiplexing, signifying parallel processing capability of multiple diffractive beams.
3. Light intensity distribution exchange, which is the exchange function of intensity distribution of emission beam through weighted intensity distribution of the incidence beam.
4. Wavelength selection, which is the wavelength-by-wavelength transmission filter function and wavelength dispersion function (refractive type and DOE type).

By making the most of the DOE attributes, such as the diffractive phenomenon and dependence on wavelength, a variety of configurations can be generated to fulfill such functions as divergence, collimation, Fourier transformation, and optical interconnection.

Applications encompass a wider range: optical measurements, optical lenses, optical sensing, spatial focal plane optics, optical interconnections, optical data stage, active optics, and

TABLE 4.2 Characteristics of the Computer-Generated Hologram (CGH) Techniques

CGH technique	Quantization	Complex data encoding	Diffraction efficiency	Fabrication file size	Fabrication	Cost of DOE
Lohman	Good	Yes	Very poor	Very large	Easy	Very low
Burch	Medium	Yes	Very poor	Large	Easy	Very low
Lee	Poor	Yes	Poor	Very large	Difficult	Medium
Kinoform	Very good	No	Very high	Small	Medium	Low
Complex kinoform	Good	Yes	Medium	Very large	Easy	Medium

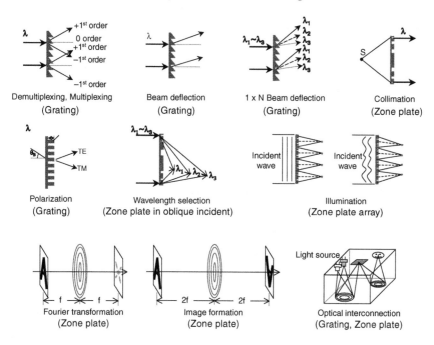

Figure 4.8 Wavefront conversion functions of diffractive optical elements.

lithographic, visual, and medical optics. Binary optics have the following merits:

1. Compact, light, and plane structure.
2. Higher efficiency by elevating approximation level.
3. Formation of imagery spot caliber approximately the same as the diffraction limit in the case of concentrating lens.
4. Flexible design.
5. Reproducibility by replica.

Table 4.3 summarizes DOE functions and application fields. Furthermore, DOEs can be distinguished between free-space and waveguide optics. In free-space optics, a light wave is not confined laterally. Rather, it is guided by lenses (as key elements in free-space optics), beam splitters, and mirrors, which are positioned at discrete positions in a longitudinal

TABLE 4.3 Functions and Application Fields for Diffractive Optical Elements (DOEs)

Elements	Functions	Application fields
Beam array generator (fan-out element)	Generates two or more beam arrays	Optical information processing, optical communication, multibeam printer, multibeam recording
Beam-shaping element	Gaussian brings change into flat-top shape, wavefront control	Laser processing, laser reaper, astronomical observation, communication
Waveguide grating	Optical coupler, splitter, wavelength selection	Optical communication, sensor, waveguide device
Planar diffractive element	Lens function, optical splitter, polarization control	Optical connection
Holographic element	Optical splitter, 3D display, optical operation, wavelength selection	Optical connection, 3D relation, HMD
CGH, SLM	Arbitrary waveface control	Optical information processing, optical connection
Element for pickup	Compound function, two focusing	Optical recording
Fiber grating	Wavelength selection	Optical communications, sensor
Hybrid-type DOE	Lens function, aberration correction	Various kinds of lens
Dispersion grating	Prism, interference filter	Measurement, projection optical system
Diffractive lens array	Beam splitter, optical coupling, optical operation	Microoptics, array illuminator, optical information processing, optical communications, interconnections
Diffuser	Control of dispersion	Display system

Note: CGH, computer-generated hologram; SLM, spatial light modulator; HMD, head mounted display.

direction (i.e., along the optical axis). In integrated optics, one can find "hybrid" structures that combine waveguide and free-space optics, but the principal emphasis is placed on the spatial focal plane. Yet another distinction has to be made between passive and active optics. By *passive optics*, we mean optical elements for light propagation, such as waveguides, lenses, lens arrays, beam splitters, and so on. By *active optics*, we mean optoelectronic devices for light generation, modulation, amplification, and detection. Application fields as described here can yield a variety of passive DOEs.

4.4 DIFFRACTION THEORY AND NUMERICAL ANALYSIS OF DIFFRACTIVE OPTICAL ELEMENTS

Regarding design and evaluation of DOEs, there are a number of theories, ranging from approximation to rigid ones. Table 4.4 divides these theories into three categories, according to geometrical optics, wave optics, and electromagnetic optics. Geometrical optics are employed for lens designs. Ray tracing is a method of approximating wavelength to zero as it proceeds through an isotropic and coherent medium. Hence, a ray of light is treated as a beam; refractive and reflective angles are calculated by the change in refractive rate according to the medium. For instance, with concentrating lenses, a ray-tracing method enables an imagery point to be calculated at the focal distance. However, wave motion will be missed in the evaluation. Wave theory is conducive for compensating for this loophole in the method. Provided a light is regarded as a scalar wave, imagery spot caliber and diffraction efficiency can be detected to account for interference and diffraction.

The principle of DOE dynamics is in accordance with the Huygens principles. It postulates that the interaction effects between periodical refractive rate distribution and the incidence electromagnetic wave should direct an emitted electromagnetic wave according to the wavelength.

Figure 4.9 classifies different approaches for the description of DOE diffraction. A record of diffractive phenomena requires better understanding of the relationship between

TABLE 4.4 Numerical Methods for Diffractive Optical Elements

	Input parameters	Output parameters	Grating period	Examples
Geometric optics	Light source Phase distribution Diameter Distance Refractive index	Beam trace Spot diagram	All areas	CODE V, ZEMAX, OSLO
Wave optics	Light source Phase distribution Diameter Distance Refractive index Phase	Phase Intensity	$\Lambda > 10 \lambda$ for wavelength λ and period Λ	Fresnel–Kirchhoff integration, angular spectrum
Electromagnetic optics	Light source Phase distribution Diameter Distance Refractive index Phase Polarization	Phase Intensity Polarization	All areas	Differential method, integral method, finite difference method

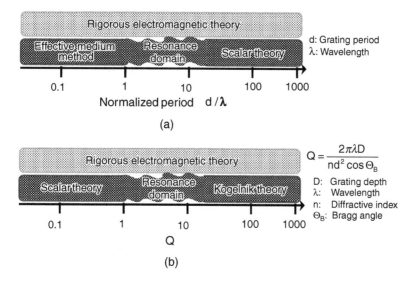

Figure 4.9   Methods to treat grating diffraction problems, classified according to the grating period and thickness: (a) problems for grating period; (b) problems for grating thickness.

grating period *d* and wavelength (λ). The rigorous diffraction theory is naturally valid for all values of *d/λ*.

As the wavelength is assumed to be comparatively and sufficiently long in its domain and thin vis-à-vis its grating period, the scalar diffraction theory can apply. Nevertheless, Figure 4.10 exhibits that as a grating period reaches a length less than ten times the wavelength, the scalar diffraction theory loses its explanatory power for explaining its traits precisely. In response, electromagnetic optical elements have to be brought back into consideration, termed *vector diffraction theory*. As the grating period becomes even thinner and reaches the level of 0.1 times the wavelength, an effective refractive rate method will be utilized to tackle the problem of average grating structure.[36] This device is called *near-field grating* and is attracting attention in research. In this domain, the vector diffraction theory is also essential in ensuring validity of the approximation.

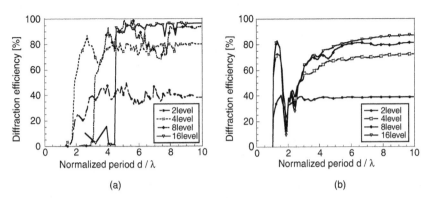

Figure 4.10 Diffraction efficiency curves for multilevel gratings as a function of the normalized period d/λ, with normal illumination from the substrate of refractive index n = 1.5 to air: (a) using the Fresnel–Kirchhoff formula; (b) using rigorous coupled-wave analysis (RCWA).

Moreover, concerning thickness D of the grating period d, parameter $Q(Q = 2\pi D/n_0 d^2)$ is known as an indicator. If $Q < 1$, thin-grating Raman–Nath diffraction proves the validity of transmission distribution approximation as multiple diffractive waves are generated. If $Q > 10$, it is called either thick grating or volume grating.[37] When refractive index modulation is negligible and the incidence angle is close to the Bragg angle, Kogelnik's two-wave, first-order, coupled-wave theory is efficient.[38] In contrast, when refractive modulation is large or has a relief-type diffraction, vector diffraction theory is employed. Last, approximation theories have not been established for the medium range (i.e., $1 < Q < 10$). Thus, as a rigorous coupled-wave analysis, the vector diffraction theory (electromagnetic theory) is required.

4.4.1 Numerical Analysis by Scalar Diffraction Theory

Scalar diffraction theory denotes numerical analysis on isotropic space based on a scalar Helmholtz's equation. My research group has employed the following methods: phase function for designing a diffractive lens and the Fresnel–Kirchhoff

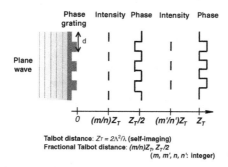

Figure 4.11 Principle of TAIL. A phase Ronchi grating is converted into amplitude gratings at fractional Talbot distances.

diffraction formula for its evaluation. However, as a numerical analysis for 2D Talbot array illuminator (TAIL), a more rigid and less-complex method is used with the angular spectrum.[39] As Figure 4.11 exemplifies, TAIL is a diffraction phenomenon near the grating period that causes a cycle of the same level of intensity distribution as the transmission rate of the grating at the Talbot distance ($Z_t = 2D/\lambda$). By lensless beam conversion, a spot array can be formed. This is an array illuminator applicable to optical interconnection.[40–42] My group has contributed to the research area by developing a new TAIL, an intensity/phase modulation type with capability for phase control.[43,44]

4.4.1.1 Fresnel–Kirchhoff Diffraction Formula

This section discusses the Fresnel–Kirchhoff diffraction formula.[45] As in Figure 4.12, on the DOE surface Q is the light source, P is the measurement point, and n is the normal vector on the sphere.

On the point P, complex amplitude U_p can be described as follows:

$$U_p = -\frac{iA}{\lambda} \int_{-\infty}^{\infty}\int_{-\infty}^{\infty} a(\rho,\theta)\frac{e^{(\rho,\theta)}e^{ikr(\rho,\theta)}}{r'(\rho,\theta)r(\rho,\theta)}b(\rho,\theta)d\rho d\theta \quad (4.23)$$

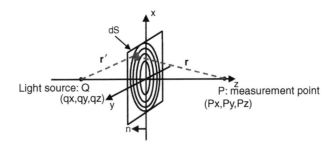

Figure 4.12 Fresnel–Kirchhoff integral diffraction formula.

where A represents the proportionality constant in regard to the incidence wave amplitude, λ is the wavelength, and $b(\rho,\ \theta)$ is the amplitude and phase modulation function of the zone plate.

The numerical analysis program for DOEs constructed by this equation is displayed in Table 4.6. The program manifests diffractive efficiency depending on the number of MLZP levels and 3D image profiles. Experimental results using fabricated MLZPs proved approximately identical to the calculated value, ascertaining the validity of scalar diffraction theory.[46]

4.4.1.2 Angular Spectrum Method

The 2D Fourier transformations for complex amplitude U_p regarding the x–y plane can be depicted as in Equation 4.23.

$$A_0(f_x, f_{y;}\, 0) = \iint\limits_{-\infty} U(x, y, 0)\exp[-i2\pi(f_x x + f_y y)]dxdy \quad (4.24)$$

Assuming U is the inverted Fourier transformation of the spectrum and is fitted into Equation 4.25, the following equation can be drawn.

Plane waves proceeding in the direction of cosines $(\alpha,\ \beta,\ \gamma)$ can be represented as in

$$U(x, y, 0) = \iint\limits_{-\infty} A_0(f_X, f_{Y;}\, 0)\exp[-i2\pi(f_X x + f_Y y)]df_X df_Y \quad (4.25)$$

$$B(x, y, 0) = \exp\left[i\frac{2\pi}{\lambda}(\alpha x + \beta y + \gamma z)\right] \quad (4.26)$$

$$\gamma = \sqrt{1 - \alpha^2 - \beta^2} \qquad (4.27)$$

$$\alpha = \lambda f_X, \ \beta = \lambda f_Y, \ \gamma = \sqrt{1 - (\lambda f_X)^2 - (\lambda f_Y)^2} \qquad (4.28)$$

In the Fourier decomposition of U, the complex example amplitude of the plane-wave component with spatial frequencies (f_x, f_y) is simply $A(f_x, f_y; 0)df_x, df_y$ evaluated as $(f_x = \alpha/\lambda, f_y = \beta/\lambda)$. For this reason, the function

$$A\left(\frac{\alpha}{\lambda}, \frac{\beta}{\lambda}; 0\right) = \int\int_{-\infty}^{\infty} U(x, y, 0) \times \exp\left[-i2\pi\left(\frac{\alpha}{\lambda} x + \frac{\beta}{\lambda} y\right)\right] dxdy \quad (4.29)$$

Equation 4.29 is called the angular spectrum of the disturbance $U(x, y, z)$.

The angular spectrum can be calculated from $\exp[i2\delta(f_X x + f_Y y)]$, where Z is equal to 0.

Furthermore, complex amplitude at the distance z from the grating can be denoted as Equation 4.29 because U needs to meet the criteria of Helmholtz's equation:

$$U(x, y, 0) = \int\int_{-\infty} A_0\left(\frac{\alpha}{\lambda}, \frac{\beta}{\lambda}, z\right) \exp\left[-i2\pi\left(\frac{\alpha}{\lambda} x + \frac{\beta}{\lambda} y\right)\right] d\frac{\alpha}{\lambda} d\frac{\beta}{\lambda} \quad (4.30)$$

Equation 4.31 is written as

$$A\left(\frac{\alpha}{\lambda}, \frac{\beta}{\lambda}; z\right) = A_0\left(\frac{\alpha}{\lambda}, \frac{\beta}{\lambda}\right) \exp\left(i\frac{2\pi}{\lambda}\sqrt{1 - \alpha^2 - \beta^2} z\right) \quad (4.31)$$

$$\nabla^2 U + k^2 U = 0 \qquad (4.32)$$

at all source free points.

Direct application of this requirement to Equation 4.31 shows that it must satisfy the differential equations

$$\frac{d^2}{dz^2} A\left(\frac{\alpha}{\lambda}, \frac{\beta}{\lambda}; z\right) + \left(\frac{2\pi}{\lambda}\right)^2 [1 - \alpha^2 - \beta^2] A\left(\frac{\alpha}{\lambda}, \frac{\beta}{\lambda}; z\right) = 0 \quad (4.33)$$

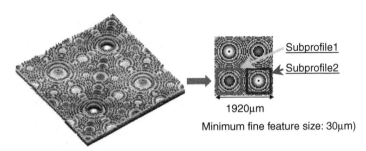

Figure 4.13 Simulation result of phase profile of the new TAIL for N_{2D} = 1024 (N_{1D} = 32), demonstrating the similarity to binary zone plates using angular spectrum.

Using the program devised by Fourier transformation and angular spectrum, configuration for a modulated TAIL was designed with intensity and phase and 32 levels as a basic unit (Figure 4.13). As the number of masks increases, errors and cost induced by complex process fabrication methods naturally aggravate efficiency. This bears a conclusive methodological prescription: The fewer phase levels there are, the better. As an application to optical interconnection, experimental results analyzing Talbot effects by a multimode slave waveguide device are shown in Figure 4.14a and Figure 4.14b. In addition, as shown in Figure 4.14a, conventional TAIL raises distortion of intensity distribution (i.e., walk-off effects) at the end of high-contrast Talbot intensity distribution because of transmission in a spatial wave plane, finite phase grating, or illumination area. In contrast, waveguide device, Talbot array illuminator (WAIL) is composed of guided wave path and phase grating placed near the aperture in Figure 4.14b. Refraction on the side of the waveguide device is conducive to generating effects of an infinitely reproduced aperture.[47]

As underlined by these two types of scalar diffraction theories, relative readiness to devise a program and conduct a numerical analysis characterizes its major merits. Its high utility peaks under certain conditions, such as a large-scale grating period (e.g., optical device larger than wave order), and in characteristic evaluation.

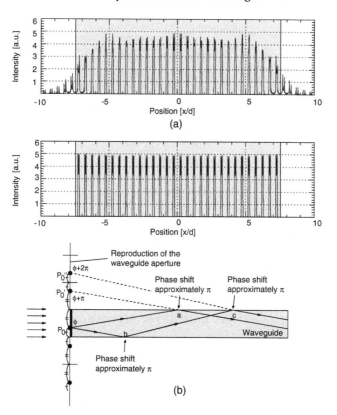

Figure 4.14 Calculated intensity distribution of TAIL and waveguide device, Talbot array illuminator (WAIL) using phase grating with compression ratio N_{1D} = 4: (a) in free space at the distance z = 5/32 z_t; (b) in optical waveguide at z = $n(5/32)z_t$ for an aperture corresponding to 15 grating periods (n = 1.46, l = 0.633 nm, d = 400 μm) the principles of waveguide TAIL.

4.4.2 Numerical Analysis Based on Rigorous Electromagnetic Wave Theory

If the grating period of DOEs is the same order of magnitude as the wavelength, the use of rigorous electromagnetic wave theory is necessary. Rigorous electromagnetic wave theory entails various methods, such as mode expansion, the integral method, Kogelnik's method, and the differential method. Among the last

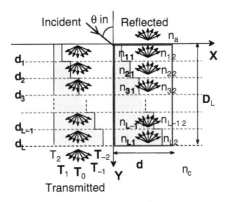

Figure 4.15   Light propagation using RCWA.

category, rigorous coupled-wave analysis[48] (RCWA) is discussed here. This method best suits the analysis of MLZPs composed of digital blaze devices, multilayer structures, and modulated refractive index gratings.

As Figure 4.15 indicates, the RCWA considers three domains: incidence/reflection, the inner grating period domain, and the transmission domain. In the case of 1D grating, coherently directed in the y-axis, a linear combination of diffractive wave applying Rayleigh expansion is used for display.

Fields U and V are defined as follows:

$$U(x,z) = \begin{cases} E_y(x,z) & (TE^*) \\ H_y(x,z) & (TM^{**}) \end{cases} \qquad (4.34)$$

$$V(x,z) = \begin{cases} -\omega\mu_0 H_x(x,z) & (TE) \\ \omega\varepsilon_0 E_x(x,z) & (TM) \end{cases} \qquad (4.35)$$

TE*: Transverse electronic mode.
TM**: Transverse magnetic mode.

Note that in the case of the TE (TM) mode, where U is the electronic (magnetic) field, V is the magnetic (electronic) field.

4.4.2.1 Incidence/Reflection Domain $(z < 0)$

The wave transmission equation in the incidence/reflection domain can be expanded by the Rayleigh process:

$$U(x,z) = I\exp[i(\beta x + a_0 z)] + \sum_{q=-\infty}^{\infty} R_q \exp[i((\beta + Kq)x - a_q z)] \quad (4.36)$$

$$V(x,z) = g_{00}^a a_0 I\exp[i(\beta x + a_0 z)]$$

$$- \sum_{p=-\infty}^{\infty} \sum_{q=-\infty}^{\infty} g_{pq}^a a_q R_q \exp[i((\beta + Kp)x - a_q z)] \quad (4.37)$$

a_q and g_{pq}^a are given as below.

$$a_q = \begin{cases} \sqrt{k_0^2 \varepsilon_a - (\beta + Kq)^2}, & \left(k_0 \sqrt{\varepsilon_a} \geq |\beta + Kq|\right) \\ i\sqrt{(\beta + Kq)^2 - k_0^2 \varepsilon_a}, & \left(k_0 \sqrt{\varepsilon_a} < |\beta + Kq|\right) \end{cases} \quad (4.38)$$

$$g_{pq}^a = \begin{cases} \delta_{pq}, & (TE) \\ \delta_{pq}/\varepsilon_a, & (TM) \end{cases} \quad (4.39)$$

where I is the transmission amplitude of qth-order light; R_q is the reflection amplitude of the qth-order light, d is the grating period, λ is the incidence wavelength, R_a is the refraction index on the incidence, θ_{in} is the incidence angle, grating vector $K = 2\pi/d$ transmission coefficient on the x-axis $\beta = Kn_a \cdot \sin\theta_{in}$, the number of waves $k = 2\pi/\lambda$, and a_q is the transmission constant on the z-axis in the incidence/reflection domain.

4.4.2.2 The *l*th Layer in the Grating $(D_{l-1} \leq z \leq D_l)$

Within the grating domain, devices for analysis are divided into the number of layers; the grating configuration in each layer can be represented by stepwise periodic function in the form of relative permittivity of dielectric constant distribution, taking into account the refractive index distribution.

$$\varepsilon(x) = \sum_{q=-\infty}^{\infty} \varepsilon_q \exp(iKqx) \quad (4.40)$$

Transmission constant b_n^l on the z-axis and the corresponding proper vector are exhibited by function rows of plane waves and known quantities. Proper vectors can be fitted as follows:

$$U_{qn}^a = \begin{cases} S_{qn}^l, & (TE) \\ P_{qn}^l, & (TM) \end{cases} \tag{4.41}$$

Hence, a general solution in the electronic field in the lth layer of the grating $(D_{l-1} \le z \le D_l)$ can be found by the following equations:

$$U(x,z) = \sum_{n=-\infty}^{\infty} \left\{ A_n^1 \exp\left[ib_n^l (z - D_l) \right] + B_n^1 \exp\left[-ib_n^l (z - D_l) \right] \right\} \tag{4.42}$$

$$\times \sum_{q=-\infty}^{\infty} U_{qn}^1 \exp\left[i(\beta + Kq)x \right]$$

$$V(x,z) = \sum_{n=-\infty}^{\infty} b_n^l \left\{ A_n^1 \exp\left[ib_n^l (z - D_l) \right] - B_n^1 \exp\left[-ib_n^l (z - D_l) \right] \right\} \tag{4.43}$$

$$\times \sum_{p=-\infty}^{\infty} \sum_{q=-\infty}^{\infty} g_{pq}^1 U_{qn}^1 \exp\left[i(\beta + Kq)x \right]$$

where D_l is the thickness of the lth layer, and g_{pq}^l is given in Equation 4.44:

$$g_{pq}^l = \begin{cases} \delta_{pq}, & (TE) \\ \delta_{pq}/\varepsilon_l, & (TM) \end{cases} \tag{4.44}$$

4.4.2.3 Transmission Domain $(z > D_l)$

A general solution in the electronic field within the transmission bandwidth can be calculated as follows:

$$U(x,z) = \sum_{q=-\infty}^{\infty} T_q \exp[i\{(\beta + Kq)x + c_q(z - D_L)\}] \tag{4.45}$$

$$V(x,z) = \sum_{p=-\infty}^{\infty} \sum_{q=-\infty}^{\infty} g_{pq}^c c_q T_q \exp[i\{(\beta + Kq)x + c_q(z - D_L)\}] \tag{4.46}$$

Here, within this band, transmission constants c_q, g_{pq}^c on the z-axis are defined as

$$c_q = \begin{cases} \sqrt{k_0^2 \varepsilon_c - (\beta + Kq)^2}, & \left(k_0 \sqrt{\varepsilon_c} \geq |\beta + Kq| \right) \\ i\sqrt{(\beta + Kq)^2 - k_0^2 \varepsilon_c}, & \left(k_0 \sqrt{\varepsilon_c} < |\beta + Kq| \right) \end{cases} \qquad (4.47)$$

$$g_{pq}^c = \begin{cases} \delta_{pq}, & (TE) \\ \delta_{pq}/\varepsilon_c, & (TM) \end{cases} \qquad (4.48)$$

A more detailed account of an algorithm based on theory can be found in Reference 49. Reference to this work clearly demonstrates the need to regard the diffractive wave as a vector wave and promotes better understanding of diffraction by solving the eigen equation. RCWA has merit in its rich resources, such as a number of computer libraries for the eigen value.

Figure 4.16a and Figure 4.16b show experimental results, applying the program to high-concentration TAIL in comparison with Fresnel–Kirchhoff's scalar diffraction theory.[50] As seen in Figure 4.16a, even though the minimum unit feature size becomes identical to the wavelength order, a spot array of similar minute size can still be produced with high contrast.

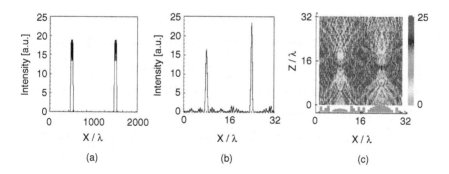

Figure 4.16 Calculated spot intensity distributions for TAIL with $N = 16$: (a) large feature size ($f_s = 60\lambda$) using angular spectrum; (b) high contrast spot for small feature size ($f_s = \lambda$) using RCWA; (c) 2D intensity distribution for (b).

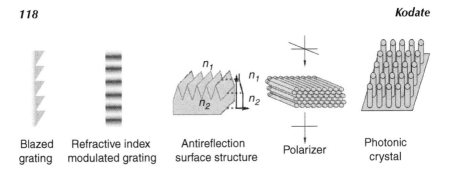

Blazed   Refractive index   Antireflection   Polarizer   Photonic
grating   modulated grating   surface structure   crystal

Figure 4.17   Examples for RCWA applied to DOEs.

Moreover, as the airy pattern is obtainable by a Fresnel lens, a spot takes a similar shape because of the evanescent wave, which cannot be detected by scalar diffraction theory. More rigid diffraction characteristics are found by this method.

Furthermore, RCWA proved its validity in analyzing various gratings, such as a photopolymer refractive index–modulated grating, a blaze grating, and a device reflection proof on an uneven surface, and a nanostructure 3D photonic crystal (Figure 4.17).

4.5 FABRICATION METHODS FOR DIFFRACTIVE OPTICAL ELEMENTS

4.5.1 Fabrication Methods for Gratings

4.5.1.1 Ruling

The classical method of fabrication was to scribe, burnish, or emboss a series of grooves on an optical surface using a diamond tool of suitable shape. Although the sag profile can be set to produce quasi-analog surface relief profiles, the grating period still remains quite large (about 20 μm) owing to the tooth size of the diamond tool. Diamond ruling is a widely used technique to fabricate blazed grating for infrared spectroscopy.

A thorough review of the history and the technology of ruling was published by Stroke[79] and a resumé by Hutley.[19]

The two main problems in ruling gratings are ensuring that the grooves are in the correct positions and that they have the correct shape. Long-term errors in the positions of the groove will give rise to aberrations in the diffracted wave front. This will reduce the image quality of the instrument and hence the resolution that can be achieved. Short-term errors will generate stray light between the diffracted orders and, if the errors are periodic, spurious peaks known as ghosts in the spectrum.

Diamond turning is the exact replica of the previous technology except that the tool no longer describes a linear but rather a circular movement. Thus, it is possible to fabricate spherical Fresnel lenses with appropriate blaze.[51] Plane surfaces as well as spherical surfaces can be machined by this technique, thus enabling the manufacture of hybrid refractive/ diffractive elements.

Evidently, these mechanical fabrication techniques have limitations in fulfilling a large fraction of the fabrication demands in diffractive optics. As device size becomes smaller, from centimeters to millimeters and micrometers, and with configuration more complex and manifold, scattering loss needs to be minimal. The fabrication process must be improved.

4.5.1.2 Optical Holography

Optical holography is a particularly important fabrication method for large gratings for spectroscopy, as described for fabricating a volume-phase holographic (VPH) grating in section 4.6.3.

The holographic recording process is based on the inter-ference of two wave fronts. This technique can provide the patterning of small feature sizes over a large area in one shot. It is usually limited to the fabrication of HOEs to be etched within the underlying substrate.

Table 4.5 shows the case of two plane wave fronts prop-agating at different angles. In the overlap area of the plane waves, an interference pattern can be fabricated consisting of linear fringes on photoresist or any other photosensitive mate-rials. The period of this grating is determined by θ:

$$d = \lambda/2\sin\theta \qquad (4.49)$$

TABLE 4.5 Fabrication Methods for Diffractive Optical Elements

E-beam direct writing	Holography and etching	Lithography and etching

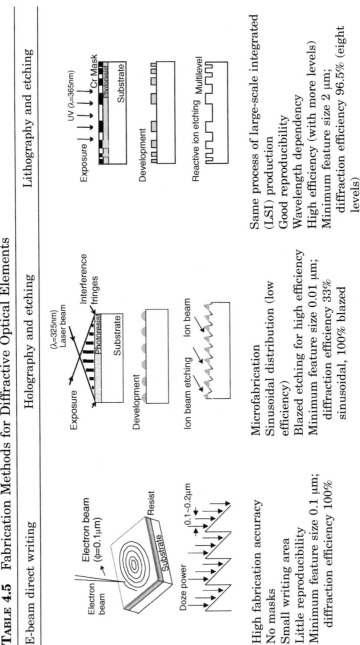
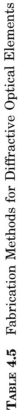

E-beam direct writing

High fabrication accuracy
No masks
Small writing area
Little reproducibility
Minimum feature size 0.1 μm;
diffraction efficiency 100%

Holography and etching

Microfabrication
Sinusoidal distribution (low efficiency)
Blazed etching for high efficiency
Minimum feature size 0.01 μm;
diffraction efficiency 33% sinusoidal, 100% blazed

Lithography and etching

Same process of large-scale integrated (LSI) production
Good reproducibility
Wavelength dependency
High efficiency (with more levels)
Minimum feature size 2 μm;
diffraction efficiency 96.5% (eight levels)

However, the relief profile here is limited to binary or sinusoidal variations, so their diffraction efficiency in a paraxial regime is less than 33.8%. To improve the diffraction efficiency, after development of the photoresist interference pattern, ion beam etching was done with the HOE rotation.

The minimum periods of HOEs are achieved for $\theta = 90°$, in which case, the fringe period is $d = \lambda/2$. Any other period can be achieved by changing the relative angle between the interfering beams or by titling the recording substrate. The holographic method has the potential for creating high submicron gratings.

More frequently, holographic recording techniques are used for the fabrication of volume grating.[52] For this purpose, thick layers ($D > 4$ μm) of photopolymers or dichromatic gelatin are used to record the interference fringes. Because of the significant improvement in quality of photopolymer materials, my group fabricated a volume phase grating with high diffraction efficiency (80% > η) using a newly developed photopolymer for astronomical observation at the 8.2-m Subaru Telescope on Mauna Kea, Hawaii,[17] as shown in Section 4.6.3 in detail.

4.5.2 Lithographic Fabrication Methods

4.5.2.1 Electron Beam and Laser Beam Writing

Direct electron beam and laser beam writing[53,54] are highly suited for the fabrication of planar continuous relief micro-DOEs, with typical microstructure of about 5-μm maximum relief and in center cases up to 10 μm or more. Writing is carried out by scanning a photoresist-coated substrate. The scanning principle is generally a combination of beam scanning over a small field and movement of an x–y stage over the writing area. The alignment of individual fields with each other is better than 5 nm in a modern machine. After development of the photoresist, a continuous relief microstructure is formed on the photoresist surface. A detailed overview of electron beam writing is discussed in relation to microlithography.[55]

Because direct writing and photolithography methods demand time and money, the replica method is now adopted. The replica method comprehends the injection molding method and slide-down and lift-up (SL) method.[56] Plane gratings for optical pickup of optical disks are in mass production by this method. Subwavelength grating is fabricated by electron beam writing and dry etching. A lithography-based method with a short-wavelength light source has also become feasible, providing further prospects for establishing an etching technique to construct structures with a high aspect ratio. To function as a subwavelength grating, the depth of a grating must be as deep as the wavelength and the types and density of plasma and temperature of the substrate must be optimized.

4.5.2.2 Microlithographic Fabrication Methods

Among the technical challenges for relief making by electron beam direct writing are dispersion effects by the beam and non-linear correction by the resist.[50] Optical interference patterns can be used to expose a photoresist layer spun upon a substrate, much like the exposure process for an optical hologram. These techniques can provide small feature size patterning over a large area in one exposure.[57]

4.5.3 Binary Optics Fabrication Methods

4.5.3.1 MLZP Fabrication Methods

With the introduction of the term *binary optics* in 1989, a new fabrication technology for DOEs was discovered owing to progress in very large scale integration (VLSI) fabrication technology. This technique was developed by Wilfried Weldkamp at the Lincoln Laboratory of MIT. With computer-generated design data and an IC fabrication technique, DOEs become practical, robust, and efficient. In particular, the binary coding of the phase quantization and fabrication sequence employed by the Lincoln researchers increased the number of phase levels formed in M process steps from $M + 1$ to 2^M. This contributed greatly to the reduction of process iterations and processing costs needed to fabricate DOEs with high diffraction efficiency.

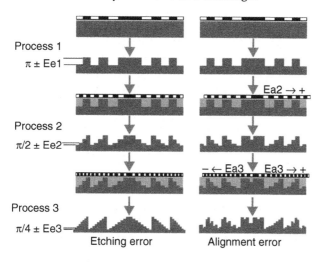

Figure 4.18 Cross-sectional view of the phase mismatch of a fabricated MLZP.

As shown in Figure 4.18, the multilayer surface relief structure of MLZPs is built up sequentially by an iterative series of lithographic and etching steps.[58] First, the MLZP pattern is designed using the scalar theory for MLZPs, and the chrome mask is fabricated by drawing it with electron beam lithography. The fabrication of an MLZP with 2^N discrete phase levels requires N mask patterns. The corresponding relief depth is given by

$$D_N = \frac{1}{2^N} \frac{\lambda}{(n-1)} \qquad (4.50)$$

where n is the refractive index of the substrate. The relation between the minimum line width W_{min}, the NA of the lens, and the level number is expressed by

$$W_{min} = \frac{\lambda}{L \times NA} \qquad (4.51)$$

Second, the photoresist is applied to the substrate. We pattern it by exposure through the mask and developing it. The pattern is then transferred to the substrate by reactive etching. Finally, the resist is removed. The fabrication of the binary

MLZP is complete. This process is repeated so that the N-level device will be manufactured.

The most important characteristic of an MLZP is its diffraction or focusing efficiency compared with the performance of the ideal lens. The diffraction efficiencies using Fresnel–Kirchhoff phenomena for MLZPs are as shown in Table 4.6. For four-level and eight-level MLZPs, 68.0% and 88.0% diffraction efficiencies are obtained, respectively. The number of phase levels increases to 16 at 93.0%.

4.5.3.2 Fabrication Errors of MLZPs

Fabrication errors occur in each of the N cycles of the fabrication process described in Section 4.5.3.1 and have a strong influence on the MLZP's diffraction performance. In exposure, position adjustment might arouse alignment error on the lateral side and etching error in depth, which are major causes for reduction of diffraction efficiency. Figure 4.19 captures this in cross section.[58]

My team and I examined the estimated effects of these errors: 10% etching error decreased efficiency by 2.4%, but there was little evidence of any effect on spot caliber. Etching errors of drying etching devices on the market are normally controlled and restricted to less than 5%, so their effects can be negligible. On the other hand, high accuracy in the alignment procedure is needed for MLZPs with high NAs, especially during the second fabrication cycle (i.e., the cycle associated with a relatively deep etching profile). Nevertheless, taking into account the currently available etching accuracy (errors less than 5%) the influence of the etching error on the diffraction efficiency and the focal intensity distribution remain small and no further consideration.

To avoid these errors, gray level mask patterns can introduce various phase levels in the photoresist during the photolithographic process.[59] Thus, it is possible to fabricate multilevel phase relief DOEs with a single mask. A single photolithographic step is followed by a single etching step. If the pattern in the mask can be encoded with sufficient dynamic range, it is possible to fabricate microelements. The dynamic range of the gray scale masks determines the achievable profiling depth.

TABLE 4.6   Diffraction Efficiencies for Various Phase Level Numbers of Diffractive Optical Elements

	4 Levels	8 Levels	16 Levels
Sectional view by atomic force microscopy			
Theoretical 3D spot profile			
Efficiency (%)	81.1	95.5	98.7
Experimental efficiency (%)	68.0	88.0	93.0

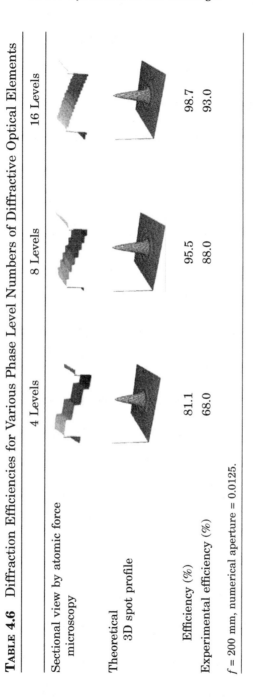

f = 200 mm, numerical aperture = 0.0125.

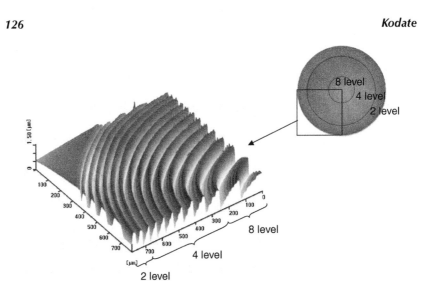

Figure 4.19 Cross-sectional view of fabricated two-, four-, and eight-level hybrid multilevel zone plates (HMLZPs).

Because of the limited dynamic range, however, a more interesting application for gray scale lithography is the fabrication of blazed DOEs.

The next section introduces a hybrid multilevel zone plate (HMLZP) design that has the potential to reduce the influence of fabrication errors and provide MLZPs with high NAs that would otherwise exceed the fabrication's resolution limit.

4.5.3.3 HMLZP Fabrication

Many applications require MLZPs that exhibit both high NA and high diffraction efficiency. To satisfy this demand, the number of levels needs to be increased and the aperture size enlarged. In such a straightforward design, however, the minimum line width of the MLZP becomes quite small because the resolution limit of fabrication for the MLZP is quickly exceeded, and the influence of fabrication errors increases. The use of an HMLZP[58] consisting of a high-level structure in the center surrounded by a low-level structure at the rim may mitigate this problem to an extent, for example, the design of an element for λ = 633 nm and f = 10 mm with a resolution limit of 1 μm in the fabrication process. A cross-sectional view of a fabricated HMLZP is shown in Figure 4.19. The highest achievable NAs of HMLZPs for an eight-level element, a four-level element, and a two-level element are 0.08, 0.16, and 0.32, respectively.[53] A combination of regions with different numbers generally results in a phase mismatch between the parts of the wave front that pass through those regions. This phase error causes reduction of focusing efficiency and therefore needs to be corrected.

In what follows, the phase error is analyzed and two methods of correction are introduced.

The nth order of the diffracted light behind a 1D binary staircase blazed grating consisting of L levels with a normalized period d is described by the nth-order Fourier coefficient, given by

$$c_n = \frac{1}{d}\int_0^\Lambda t(x)\exp\left(-i\frac{2\pi}{d}nx\right)dx \qquad (4.52)$$

where $t(x)$ denotes the grating's complex transmittance. The complex amplitude of the first order, which is the relevant order for the focal distribution of an MLZP, is given by

$$c_1 = \exp\left(-i\frac{\pi}{L}nx\right)\sin c\left(\frac{1}{L}\right) \qquad (4.53)$$

Hence, the phase difference between zones of levels L_1 and L_2 becomes

$$\phi = \frac{\pi}{L_1} - \frac{\pi}{L_2} \qquad (4.54)$$

One example shows that the phase mismatch for a two- and four-level combination amounts to $\pi/4$ and for a four- and eight-level combination amounts to $\pi/8$. As depicted in Figure 4.20, we can easily adjust such a phase mismatch either by employing Lohmann's detour phase principle[60] (i.e., slightly shifting the structure in the lateral direction) or by adjusting the etching depth. Table 4.7 summarizes both numerical and experimental results for MLZPs designed and fabricated according to these two correction methods. In the case of depth adjustment, an additional fabrication process should be introduced, which will risk the increase of fabrication errors. We eventually decided to use the correction method that

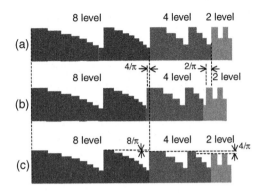

Figure 4.20   Two methods for compensating for the phase mismatch of HMLZPs: (a) no correction; (b) correction by means of a lateral shift; (c) correction by a phase depth adjustment.

employs a lateral shift and analyzed it with respect to mask alignment errors.

Alignment errors are common to each fabrication process and can hardly be avoided. In what follows, HMLZPs are examined, including alignment errors, and the condition for the optimum combination of zones with different levels is determined. Taking into account a presumed mask alignment error, Figure 4.21 shows the flow chart of optimum design for a given fabrication accuracy.[50]

TABLE 4.7   Calculated Results of the Diffraction Efficiency for a Hybrid Multilevel Zone Plate with and without Phase-Mismatch Correction for a NA of 0.02

Level	Correction	Diffraction efficiency (%)	
		Measured	Calculated
2	None	40.6	40.5
2 and 4	None	58.0	59.8
	In radial direction	64.4	67.7
	In depth direction	61.4	67.2
4 and 8	None	85.5	93.0
	In radial direction	87.5	94.5
	In depth direction	87.4	94.5
8	None	93.2	94.9

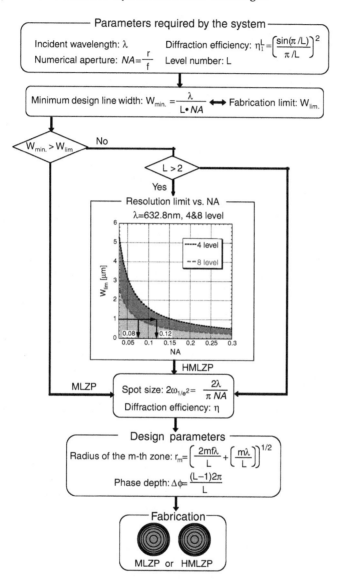

Figure 4.21 Design procedure and choice between either MLZPs or HMLZPs, depending on the fabrication resolution limit and the desired level number.

4.6 APPLICATIONS OF DIFFRACTIVE OPTICAL ELEMENTS

As noted, DOEs have been regarded as one of the most prospective devices, possessing excellent function and features. Table 4.3 presents a proposed device and its applications and functions[61] and introduces application examples and future possible applications.

4.6.1 Optical Lenses

With high demand for microoptical elements in optical design and fabrication, DOEs have already acquired recognition as prospective and significant devices. The system includes the following tools: microscopic lenses integrated with refractive optical elements,[62] hybrid lenses with two focal points for compact disc (CD) and digital video disc (DVD) use,[63] super-compact infrared sensors with a diffractive microlens,[64] and stacked DOEs for taking a photograph.[65] Figure 4.22 shows the multilayer DOE for camera lenses; it arranges two single-layer lenses to remove unnecessary diffractive light and make the most of all incident light as photographic light. By combining a multilayer DOE and refractive optical element within the same optical system, chromatic aberration (color smearing),

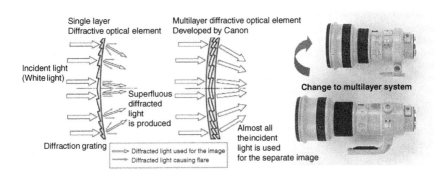

Figure 4.22 DOEs for camera lenses: diffraction properties of a single-layer and a multilayer DOE; EF 400 mm f/4 DOEs USM built-in multilayer.

which adversely affects image quality, can be corrected even more effectively than with a fluorite element.

Moreover, to test the light-converging function of a spherical lens, a number of small electrodes possessing a wheel structure on a concentric circle are combined. By external input pressure, the refractive index of the material is altered. Likewise, a variable focal liquid crystal (LC) microlens is also consolidated into an LC projector.[66]

4.6.2   Optical Sensing and Wavelength Division Demultiplexing

The bandwidth explosion in the optical telecommunications field will inevitably cause enormous problems for densification of wavelength division demultiplexing (WDM) optical networks. Bragg gratings will be a key technology for their characteristics. They consist of gratings formed in the core of doped single-mode optical fibers when exposed to a periodic pattern of ultraviolet (UV) light. The grating is physically inscribed as an index modulation within the fiber. Bragg gratings are also essential for fiber sensing[67] and wavelength demultiplexing because the fiber gratings can act as an internal "mirror" or ultranarrow rejection band for a specific wavelength and leave the other wavelengths quasi unperturbation. Investors and analysts worldwide predict that Bragg fiber grating will be one of the major DOE markets within the next 5 years.

4.6.3   High-Dispersion VPH Gratings for Astronomical Observation

Dispersion elements are essential in spectroscopic observation to detect feeble light efficiently in dark and far-off galaxies several billion light years away. Spectra (OH [hydroxyl] airglow lines) from OH radical excited by UV rays from the sun emit light 100 to 1000 times as bright as that of a feeble celestial body, which serves as a serious obstacle for accurate observation. The spectrograph currently used for astronomical observation at the 8.2-m Subaru Telescope on Mauna Kea, Hawaii, contains a grism that consists of a relief transmission grating

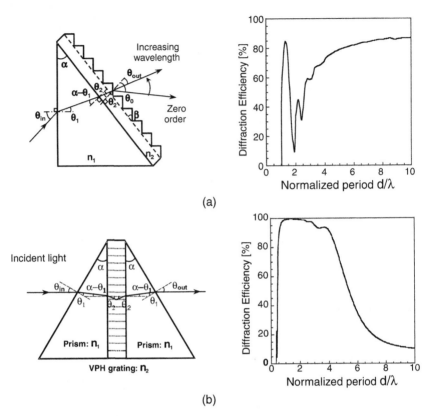

(a)

(b)

Figure 4.23 Geometry of two types of grism and diffraction efficiency curves vs. normalized period for two grisms. α, appendix angle
of prism; β, appendix angle of grating; θ_{in}, incident angle; θ_{out}, first-
order angle; n_1, refractive index of prism; n_2, refractive index of
grating. (a) blazed grism; (b) VPH grism.

attached to a prism[68] (Figure 4.23a). This configuration, however, limits the possibilities for improvements to transmission
efficiency and spectral resolution.

My group designed and fabricated a new type of grism by
sandwiching a thick VPH grating between two prisms (Figure
4.23b).[17] The process of looking for the optimum parameters
such as grating thickness and strength of the refractive index
modulation is crucial in obtaining a well-performing grism. The
refractive index distribution of the VPH grating varies in a

nearly sinusoidal way. RCWA (section 4.4.2) is well suited to analyze such a grism and to pinpoint problems that may arise during the fabrication process and adversely affect the overall efficiency of the spectroscopic measurement.

The VPH grism was constructed with a 430-nm incident wavelength, and it was found that high-distribution diffraction ($R = 2500$) in a visible light domain can be carried in the Subaru Telescope. At the optimal conditions of 430-nm incident wavelength and 1.0-μm grating period, it was concluded that, at the 10-μm grating thickness, the refractive index modulation of 0.02 was the highest efficiency (~90%) by Bragg angle incidence of $\lambda = 430$ nm.[17]

4.6.4 High-Compression TAILs

The TAIL is well known for its capability to transform a monochromatic light wave efficiently into an array of bright spots in the near or Fresnel field behind phase gratings. The illumination planes lie at fractions of the Talbot distance $Z_t = 2d^2/\lambda$, where d represents the grating period, and λ is the wavelength of the illumination. TAILs achieve an almost-lossless transformation of a plane wave into a regular line or spot intensity pattern.[69] Since Lohmann first described the working principle of the TAIL in the late 1980s, many articles have been published dealing with their analysis and synthesis.[43]

Owing to rapid progress in the field of binary optics, DOEs can now be fabricated at moderate cost and with high precision. The capability of providing both binary and multilevel grating profiles could finally lead to a wider acceptance of the Talbot grating as an efficient array illuminator (Figure 4.24). When using the LC panel as a pure phase or wave front modulator (e.g., for beam deflection or focusing), diffraction effects deliberately introduced by the first Talbot plate cannot be corrected further by a static second plate. As shown in Figure 4.24, this problem is solved by combining only one Talbot plate generating an intensity pattern with a duty cycle of 80% and an LC phase modulator with large active areas. Based on the Fresnel–Kirchhoff diffraction formula, both near and far field of Talbot plates with different duty cycles were

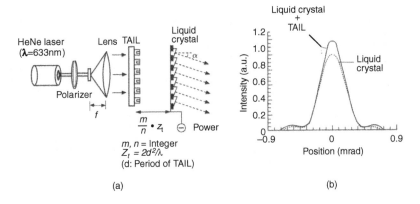

(a)   (b)

Figure 4.24 (a) Optical setup for the far-field measurement of a beam-steering experiment using a TAIL; (b) calculated far-field intensity of the main lode.

calculated, and the efficiency of the overall system was maximized by paying special attention to reduce both the loss in the inactive regions of the LC panel and the intensity of the light diffracted into higher orders to a minimum.

To obtain the TAIL's best illumination performance, it is necessary to optimize the efficiency and the compression ratio regarding the fabrication cost (i.e., the number of required lithographic masks). In this respect, the features, performance, and the illumination efficiency of two families of 2D TAILs have been discussed (Figure 4.13).[42]

4.6.5 Free-Space All-Optical Demultiplexing Module

Because it is free from electrical parasitics, the use of the all-optical switch array is an attractive approach for achieving high-speed signal processing for optical communication and interconnection beyond the gigabit-per-second range. However, two incident beams and one outgoing beam for each switch may require complex geometry; thus, the module assembly procedure remains a serious obstacle for its realization. My group proposed a new geometry that aims at the realization

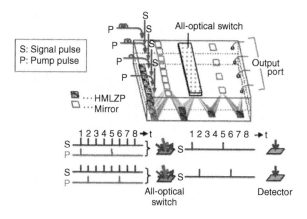

Figure 4.25 Concept of an all-optical switching module composed of a glass optical platform equipped with DOEs.

of a compact and robust all-optical time division demultiplexing module by combining the suitable absorber switch approach with diffractive optical technology.[70] A free-space microoptical platform employing DOEs (off-axis MLZPs) provides a noncollinear beam combination as well as focusing function for signal and pump beam on the saturable absorber switch array.

Figure 4.25 illustrates the concept of the all-optical time division demultiplexing module. A high-bit-rate signal stream is separated into multiple channels with different optical time delays. Optical pump pulses are synchronized with the signal pulses by a clock extraction optical circuit. Both the signal beams and the pump beams are fed to the optical platform with a fiber array placed perpendicular to the optical platform. An all-optical saturable absorber switch array is attached to the glass platform with suitable glue. The switched signals are either taken out of the platform with fiber couplers or directly detected by a photodiode array.

A 5-mm platform thickness d was chosen in consideration of the limitations of the processing equipment and of the desired compactness. The aperture size D of the first lens and thus also the fiber array spacing was set to 850 μm.

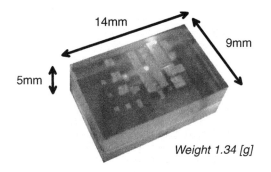

Figure 4.26 Photograph of the fabricated all-optical demultiplexing module.

Different from other planar integrated free-space optical systems demonstrated so far, this design uses diffractive optical systems on both sides of the platform. In this work, eight-level, off-axis HMLZPs were fabricated on a 2.5-mm thick fused silica glass substrate in three subsequent photolithography and reactive ion etching processes. The etching depths and their deviations from the design values measured 268 nm (–0.3 nm), 133 nm (–1.2 nm), and 69 nm (+1.9 nm) for the two-, four-, and eight-level patterns, respectively. The minimum lateral feature size given by technological limitations in the fabrication process was 2 μm. Bottom and top parts were aligned with the help of alignment marks on the rear side and were attached by a UV-curable adhesive. Figure 4.26 shows a photograph of an optical platform. A preliminary switching experiment for time division multiplexing–demultiplexing was performed using 25-psec gained–switched LDs, and switching time around 200 psec was confirmed. Moreover, a wave theoretical design procedure was formulated, applied to the analysis of the first design, and used for a new design of all-optical switching modules. Assuming the zone plate fabrication technology with a minimum feature size of 1 μm, the spot size of 5 μm is predicted to be achievable, while ensuring 85% diffraction efficiency. With these improved performances, the present design procedure is a favorable approach to realizing the all-optical switching module operating at λ = 1.55 μm.[19]

4.6.6 Multilevel Zone Plate Array for a Compact Optical Parallel Joint Transform Correlator Applied to Facial Recognition

With the progress of information technology, the need for an accurate personal identification system based on biological characteristics is increasing the demand for this type of security technology instead of conventional systems using identification cards or PINs (personal identification numbers). Among other features, the face is the most familiar element and is less subject to psychological resistance. In contrast to digital recognition, optical analog operations process 2D images instantaneously in parallel using an optical Fourier transform function. Two methods were proposed in the 1960s: the VanderLugt correlator[70] and the joint transform correlator (JTC).[71]

My group proposed a new scheme using a multichannel parallel joint transform correlator (PJTC) to make better use of spatial parallelism using a multilevel zone plate array (MLZPA) to extend a single-channel JTC. The PJTC scheme was applied to facial recognition by adding the simple and powerful pre- and postdigital processing techniques. Taking advantage of compact and efficient MLZPAs (an eight-level binary optical element with diffraction efficiency above 70%), two compact optical parallel correlators (COPaC II, patented by COPaC, $20 \times 24 \times 43$ cm^3, 6 kg, 6.6 faces/sec throughput time) were designed, assembled, and tested.[18,72,73]

Using the PJTC, each zone plate acts as a Fourier transform lens for an independent channel. Therefore, an MLZPA produces an array of joint power spectrums of individual joint pairs without lateral overlapping (Figure 4.27). The facial images under test and the reference facial images are initially stored in a personal computer. The transfer of the photographic images from the computer to the electronically addressed spatial light modulator (ESLM) would limit the throughput if no data compression were applied. The principle architecture of the PJTC is shown in Figure 4.28. In an experimental evaluation of the system by one-to-one correlation using 300 front facial images, the false match and false nonmatch rates were less than 1%. The results indicated a recognition rate of greater

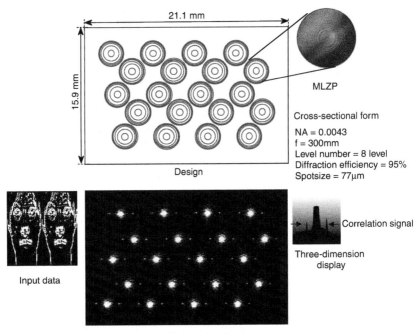

Figure 4.27 Photograph of a 20-channel (MLZPA) for Fourier transform lenses and experimental results of 20 self-correlation signals when the same facial images were input.

than 90% over a 6-month period. Facial recognition was verified using facial images obtained under arbitrary lighting conditions of 30 lx or higher without significant changes in facial expression or differences in accessories (e.g., glasses), as shown in Figure 4.29. Optical facial recognition is therefore highly applicable in a range of systems, including security systems.[74]

Use of an LC optical modulator in the PJTC correlator is essential, but the throughput is limited by the response of the spatial light modulator (SLM). Experiments are currently under way to solve this issue by applying a VanderLugt correlator based on the same algorithm. A fully automatic ultra-high-speed system capable of processing 1000 faces/sec by 1 channel has been successfully developed (Figure 4.30).[75]

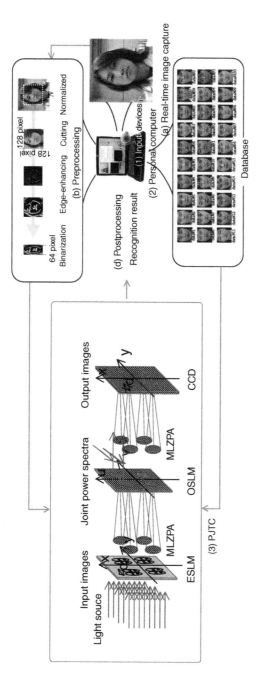

Figure 4.28 Algorithm for face recognition COPaC; OSLM, optical addressed spatial light modulator; CCD, charge-coupled device.

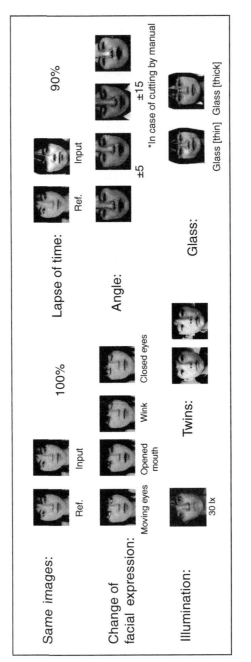

Figure 4.29 Input face image example of recognition.

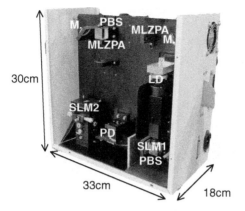

Figure 4.30 Photograph of the fabricated Fast Face-Recognition Optical Correlator (FARCO).

4.7 RECENT DEVELOPMENTS IN DIFFRACTIVE OPTICS

Optics and optical techniques will play a major role in the information society. Optical devices particularly will be developed and improved according to new and stricter demands. Larger capacities, multiple functions, and high performance are major concerns because current attention is restricted to the control of wave planes, amplitude, wavelength, and polarization. This attention should be extended to a wider range of techniques and developments, both temporal and spatial. Examples of this broader range of goals include expansion of usable wavelength bandwidth and devices and the establishment of more effective fabrication processes in the subwavelength domain.

Table 4.8 summarizes the characteristics of passive and active optical devices, the latter representing functional devices such as DOEs. Increased rise in dimensionality and improved performance realized by DOEs further promotes the demand for highly sophisticated fabrication techniques. By artificially creating multidimensional periodic structures on a crystallization level and utilizing the inherent vector and photon modes, the characteristics of wave transmission can be

TABLE 4.8 Characteristics of Passive and Active Optical Devices

Name	DOE (amplitude type)	Passive optical elements DOE (phase type)	Photonic crystal
Shape	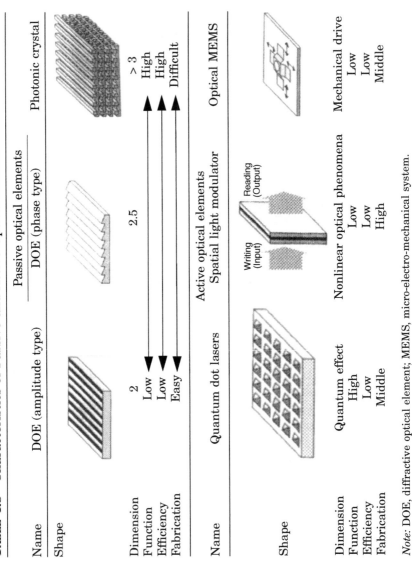		
Dimension	2	2.5	> 3
Function	Low		High
Efficiency	Low		High
Fabrication	Easy		Difficult

Name	Quantum dot lasers	Active optical elements Spatial light modulator	Optical MEMS
Shape		Writing (Input) / Reading (Output)	
	Quantum effect	Nonlinear optical phenomena	Mechanical drive
Dimension	High	Low	Low
Function	Low	Low	Low
Efficiency	Middle	High	Middle

Note: DOE, diffractive optical element; MEMS, micro-electro-mechanical system.

reproduced at even shorter lengths compared to the operating wavelength. Devices based on this concept have in fact been developed and are known as photonic crystals.[76] Detailed characterization of these devices will be necessary to make full use of vector diffraction theory in the design and evaluation of fine structures. The development of improved fabrication processes poses further challenges, with design and fabrication of nanostructures and fabrication techniques using quantum beams examples of such development. For improved dynamic control, electro-optical effects such as crystallization are more frequently used, and spatial light modulators have been examined for control.[77]

The optical parallel correlator used for facial recognition discussed in the preceding section includes a voltage excitation DOE for a real-time display of images, as an optical filter, and as a photowriting SLM. However, these elements have specific shortcomings and need to be improved. In particular, their control capacity is limited to a phase gap of a specific wavelength, the resolution power is limited by the thickness of the electro-optical materials, and their use is restricted to one polarization of light.

Optical microelectromechanical systems, representing fine mechanical driving circuits fabricated by micromachining techniques, are free from this dependence on polarization. Quantum dot lasers also are not affected by temperature, modulation, or threshold values. These light-emitting devices are devised by locking electrons into a 3D microscopic domain and are promising microdevices.[78]

All of these devices are prominent in that they have outstanding features and functions and the potential to provide significant developments and improvements. Through the hybridization of passive and positive devices, it is hoped that new electro-optic devices with compound functions will be realized in the near future.

This review shows clearly how this field has developed and emphasizes my group's interest and contribution. From the early days of research into microdiffractive optics in the 1970s, the application of these technologies to optical information processing is finally realized.

ACKNOWLEDGMENTS

This chapter is first and foremost dedicated to the late Emeritus Prof. M. Kamiyama, who guided me to the diffractive optics field of research. I also would like to express heartfelt thanks to Prof. T. Kamiya at the National Institution for Academic Degrees and University Education, the Ministry of Education and Technology, who has generously devoted his enthusiasm and support to our collaboration over a number of years. Many thanks to our research partners, one of whom is Dr. Werner Klaus, chief researcher at the Communications Research Laboratory, who is an expert on Talbot research. In addition, this chapter refers to a number of articles and thus is based on research done by the graduates from our laboratory. I would like to express thanks to Y. Komai, K. Oka, and other Kodate Laboratory students at the faculty of Sciences, Japan Women's University.

REFERENCES

1. *Optoelectronics Technology Roadmaps*, Optoelectronic Industry and Technology Development Association, Tokyo, Japan, 2001.

2. S. Sinzinger and J. Jahans, *Micro Optics*, Wiley-VCH, Weinheim, Germany, 1999.

3. B. Kress and P. Meyrueis, "Design and simulation of diffractive optical elements." B. Kress and P. Meyrueis, *Digital Diffractive Optics*, Wiley-VCH, Weinheim, Germany, 2000, pp. 17–129.

4. H.P. Herzig, "Design of Refractive and Diffractive Microoptics." H.P. Herzig, *Microoptics*, Taylor and Francis, London, 1997, pp. 1–52.

5. M. Kuittinen, J. Turunen, and P. Vahimaa, "Subwavelength-Structured Elements." J. Turnen and F. Wyrowski, Eds., *Diffractive Optics for Industrial and Commercial Applications*, Akademie Verlag, Berlin, 1997, pp. 303–324.

6. J.R. Legar, M. Holz, G.J. Swanson, and W.B. Veldkamp, Coherent laser beam addition: an application binary optics technology, *Lincoln Lab. J.*, 1, 255 (1988).

7. Y. Ono, Design and applications of optical system using diffractive optics, *Jpn. J. Optics*, 24, 713 (1995).

8. K. Kodate, *Handbook of Optical Computing*, Asakura, Tokyo, 1997, pp. 442–453.

9. K. Kodate, T. Kamiya, H. Takenaka, and H. Yanai, Construction of photolithographic phase grating using the Fourier image effect, *Appl. Opt.*, 14, 522 (1975).

10. K. Kodate, H. Takenaka, and T. Kamiya, Fabrication of high numerical aperture zone plate using deep ultraviolet lithography, *Appl. Opt.*, 23, 504 (1984).

11. K. Kodate, Y. Okada, and T. Kamiya, A blazed grating fabricated by synchrotron radiation lithography, *Jpn. J. Appl. Phys.*, 25, 822 (1986).

12. G.J. Swanson and W.B. Veldkamp, Diffractive optical elements for use in infrared system, *Opt. Eng.*, 28, 605 (1989).

13. K. Kodate, Y. Orihara, and W. Klaus, Toward the optimal design of binary optical elements with different phase levels using a method of phase mismatch correction, Technical Digest Diffractive Optics and Microoptics, Diffractive Optics and Microoptics 2000 of OSA Trends in Optics and Photonics Series, 2000.

14. K. Oka, A. Yamada, N. Ebizuka, T. Teranishi, M. Kawabata, and K. Kodate, Optimization of a volume phase holographic grism for astronomical observation using the photopolymer by the rigorous coupled wave analysis, proceedings of the 28th Kougaku (in Japanese) Symposium, Japan, 2003.

15. Y. Komai, K. Kodate, K. Okamoto, and T. Kamiya, Novel application of arrayed waveguide grating to compact planar spectroscopic sensors, Ninth Microoptics Conference Technical Digest, 2003.

16. Y. Komai, K. Kodate, and T. Kamiya, Improved usage of binary diffractive optical elements in ultrafast all-optical switching modules, *Jpn. J. Appl. Phys.*, 41, 4831 (2002).

17. K. Oka, A. Yamada, Y. Komai, E. Watanabe, N. Ebizuka, T. Teranishi, M. Kawabata, and K. Kodate, Optimization of a volume phase holographic grism for astronomical observation using the photopolymer, *SPIE*, 5005, 8 (2003).

18. K. Kodate, R. Inaba, E. Watanabe, and T. Kamiya, Facial recognition by compact parallel optical correlator, *Meas. Sci. Technol.*, 13, 1756 (2002).

19. M.C. Hutley, *Diffraction Grating*, Academic Press, London, 1982.

20. D. Rittenhouse, Solution of an optical problem proposed by Mr. Hopkinson, *Trans. Am. Philos. Soc.*, 2, 201 (1786).

21. J.N. Fraunhoffer, Neue modification des Lichtes durch gegenseitig Einwirkung und Beugung der Strahlend, und Gesetze desselben, Denkeshr, *Denkeshr. Kg. Akad. Wiss. Munchen*, (1821).

22. H.A. Rowland, Preliminary notice of the results accomplished in the manufacture and theory of grating for optical purpose, *Philos. Mag.*, 13, 469 (1882).

23. R.W. Wood, A remarkable case of uneven distribution of light in a diffraction grating spectrum, *Philos. Mag.*, 4, 396 (1902).

24. E. Wolf, "Progress in Optics." H. Nishihara and T. Suhara, Micro Fresnel lenses, in *Progress in Optics 24*, North-Holland, Amsterdam, 1987, pp. 1–38.

25. G.J. Swanson, Binary optics technology: theoretical limits on the diffraction efficiency of multilevel diffractive optical elements, *MIT Tech. Rep.*, 914, 1 (1991).

26. S. Sinzinger and J. Jahns, *Microoptics*, Wiley-VCH, Weinheim, 2003, pp. 3–83.

27. D. Gabor, A new microscopic principle, *Nature*, 161, 777 (1948).

28. P. Hariharan, *Optical Holography*, Cambridge University Press, Cambridge, U.K., 1996, pp. 1–66.

29. E.N. Leith and J. Upatnieks, Wavefront reconstruction with diffused illumination and three-dimensional objects, *J. Opt. Soc. Am.*, 54, 1295 (1964).

30. J.M. Lerner and A. Thevenon, *Handbook: Diffraction Gratings. Ruled and Holographics*, Jobin-Yvon, Inc., New Jersey, (1973).

31. K. Moriwaki, H. Aritome, and S. Nanba, Fabrication of submicron structures by ion beam lithography, *Jpn. J. Appl. Phys.*, 20(Suppl.), 1305 (1981).

32. Y.N. Denisyuk, Photographic reconstruction of the optical properties of an object in its own scattered radiation field, *Sov. Phys.*, 7, 543 (1962).

33. W.H. Lee, Binary computer-generated holograms, *Appl. Optics*, 18, 3661 (1979).

34. Y. Ono and N. Nishida, Holographic laser scanners using generalized zone plates, *Appl. Opt.*, 21, 4542 (1982).

35. J. Turunen and F. Wyrowski, *Diffractive Optics for Industrial and Commercial Applications,* Akademie Verlag, Berlin, 1997, pp. 1–80.

36. J. Turunen, *MicroOptics,* Taylor and Francis, London, 1997, pp. 31–52.

37. E.B. Ramberg, The hologram—properties and applications, *RCA Rev.,* 27, 467 (1966).

38. H. Kogelnik, Coupled wave theory for thick hologram gratings, *Bell Sys. Tech. J.,* 48, 2909 (1969).

39. J.W. Goodman, *Introduction to Fourier Optics,* McGraw-Hill, New York, 1996, pp. 55–62.

40. H.F. Talbot, LXXVI. Facts relating to optical science, No. IV, *London Edinburgh Philos. Mag. J. Sci.,* 19 (1836).

41. A.W. Lohman, An array illuminator based on the Talbot-effect, *Optik,* 79, 41 (1988).

42. V. Arrizon, Array illuminator with an arrangement of binary phase gratings, *Opt. Lett.,* 18, 1205 (1993).

43. W. Klaus, Y. Arimoto, and K. Kodate, Talbot array illuminators providing intensity and phase modulation, *J. Opt. Soc. Am.,* A14, 1092 (1997).

44. W. Klaus, K. Kodate, and Y. Arimoto, High-performance Talbot array illuminators, *Appl. Opt.,* 37, 4357 (1998).

45. M. Born and E. Wolf, *Principle of Optic,* Pergamon Press, Oxford, U.K., pp. 370–458 (1980).

46. A. Okazaki, W. Klaus, and K. Kodate, Design of optimized binary optical element by combining various phase levels, *Optoelectronics Laser,* 9(Suppl.), 356 (1998).

47. T. Nakayama, W. Klaus, and K. Kodate, Two-dimensional waveguide Talbot array illuminator, *J. Jpn. Women's Univ. Fac. Sci.,* 9, 25 (2002).

48. M.G. Moharam and T.K. Gaylord, Rigorous coupled-wave analysis of planar grating diffraction, *J. Opt. Soc. Am.,* 71, 811 (1981).

49. K. Oka, W. Klaus, and K. Kodate, Analysis of diffractive optical element with wavelength region using rigorous coupled-wave theory, *J. Jpn. Women's Univ. Fac. Sci.,* 10, 99 (2002).

50. W. Klaus, K. Oka, M. Fujino, and K. Kodate, Analysis of near-field intensity distributions of high-compression Talbot array illuminators using rigorous diffraction theory, *Opt. Rev.*, 8, 271 (2001).

51. M.C. Hultley, Blazed interference diffraction grating for the ultra-violet, *Opt. Acta*, 22, 1 (1975).

52. L. Solymar and D.J. Cooke, *Volume Holographic and Volume Gratings*, Academic Press, London, 1981.

53. A.H. Finester, Properties and fabrication of micro Fresnel zone plates, *Appl. Opt.*, 12, 1698 (1973).

54. M. Haruna, M. Takahashi, K. Wakabayashi, and H. Nishiyama, Laser beam lithographed micro Fresnel lenses, *Appl. Opt.*, 29, 5120 (1990).

55. M. Kufner and S. Kufner, *Microoptics and Lithography*, VUB Press, Brussels, 1997, pp. 23–80.

56. M. Ishizuka and Y. Matsuo, SL method for computing a near-optimal solution using linear and non-linear programming in cost-based hypothetical reasoning, *Proc. PRICAI'98*, 611 (1998).

57. H.P. Herzig, *Microoptics*, Taylor and Francis, London, 1997, pp. 37–84.

58. Y. Orihara, W. Klaus, M. Fujino, and K. Kodate, Optimization and application of hybrid-level binary zone plates, *Appl. Opt.*, 40, 5877 (2001).

59. M.R. Wang and H. Su, Laser direct-write gray-level mask and one-step etching for diffractive microlens fabrication, *Appl. Opt.*, 37, 7568 (1998).

60. B.R. Brown and A.W. Lohmann, Complex spatial filtering with binary masks, *Appl. Opt.*, 5, 967 (1996).

61. New Energy and Industrial Technology Development Organization, *Optical Functional Integrated System Technology*, New Energy and Industrial Technology Development Organization, Tokyo, Japan, (2001).

62. H. Kikuta and K. Iwata, Current topics in diffractive optics, *IEICE Trans. Electron.* (Jpn. ed.), C83, 173 (2000).

63. K. Maruyama, *"Introduction to Diffraction Optical Elements,"* Optronics, Tokyo, Japan 1997, pp. 30–50.

64. T. Shiono, M. Kitagawa, K. Setune, and T. Mitsuyu, Reflection micro-Fresnel lenses and their use in an integrated focus sensor, *Appl. Opt.*, 28, 3434 (1989).

65. Canon develops world's first Multi-Layer Diffractive Optical Element for camera lenses, http://www.canon.com/do-info/.

66. N. Hashimoto, K. Ogawa, S. Morokawa, and K. Kodate, Real-time optical correlation using LCTV-SLM and FZP, *IEEE Denshi Tokyo*, 32, 67 (1994).

67. K.O. Hill, Fiber Bragg grating technology fundamentals and overview, *J. Lightwave Technol.*, 15, 1263 (1997).

68. N. Ebizuka, M. Iye, T. Sasaki, and T. Sasaki, Optically aniso-tropic crystalline grisms for astronomical spectrographs, *Appl. Opt.*, 37, 1236 (1998).

69. K. Kodate, W. Klaus, N. Kanamori, S. Morokawa, and T. Kamiya, Theoretical and experimental evaluation of array illuminators for large aperture phase-only spatial light modulators, *SPIE*, 2577, 28 (1995).

70. A. VanderLugt, Signal detection by complex spatial filtering, *IEEE Trans. Inf. Theory*, 10, 139 (1964).

71. C.S. Weaver and J.W. Goodman, A technique for optically con-volving two things, *Appl. Opt.*, 5, 1248 (1966).

72. K. Kodate, A. Hashimoto, and R. Tapliya, Binary zone-plate array for a parallel joint transform correlator applied to face recognition, *Appl. Opt.*, 3, 400 (1999).

73. K. Kodate, E. Watanabe, and R. Inaba, Optoelectronic face rec-ognition system using diffractive optical elements: design and evaluation of compact parallel joint transform correlator (COPaC), *SPIE*, 4455, 42 (2001).

74. R. Inaba, E. Watanabe, and K. Kodate, Security applications of optical face recognition system: access control in e-learning, *Opt. Rev.*, 10, 255 (2003).

75. E. Watanabe and K. Kodate, Fabrication and evaluation of ultra-fast facial recognition system, Frontiers in Optics the 87th OSA Annual Meeting Laser Science 19, Tucson, Arizona 2003.

76. S. Noda, K. Tomoda, N. Yamamoto, and A. Chulinan, Full three-dimensional photonic bandgap crystals at near-infrared wave-lengths, *Science*, 289, 604 (2000).

77. S. Fukushima, T. Kurokawa, S. Matsuo, and H. Kozawaguchi, Bistable spatial light modulator using a ferroelectric liquid crystal, *Opt. Lett.*, 15, 285 (1990).

78. Y. Arakawa and H. Sasaki, Multidimensional quantum well laser and temperature dependence of its threshold current, *Appl. Phys. Lett.*, 40, 939 (1982).

79. A. Marechal and G.W. Stroke, Sur l'origine de effets de polarisation et de diffraction dans les reseaux optiques, *Proc. SPIE*, MS 83, 561 (1992).

5

Planar Optics

HIRONORI SASAKI

CONTENTS

5.1 INTRODUCTION

Massively parallel optical interconnections (Figure 5.1) are easily realized by imaging between an array of light sources and an array of photodetectors.[1,2] Because the propagating signal bandwidth is only affected by the capacitance of both light source and photodetector, ultrafast optical signal transmission is expected. In contrast, for electronics interconnections, the velocity of signal transmission depends on the capacitance per unit interconnection length. Therefore, the transmission bandwidth decreases as the interconnection length increases.

Free-space optical interconnection has the following additional advantages. Because optical beams can freely intersect without causing any mutual interference, dense interconnections are easily realized by making many optical signals propagate within a shared optical propagating medium without the degradation of signal integrity. Free-space optical interconnections are also free from the planar wiring topology constraints. Because one can fully utilize the three-dimensional (3D) volume of free space for interconnections, corresponding interconnection density is larger than that of waveguide approaches.[3]

However, free-space optical interconnections are not without problems. Because light sources, photodetectors, and

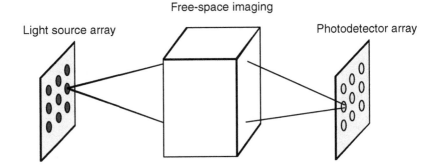

Figure 5.1 Concept of free-space optical interconnections based on the imaging between the array of light sources and the array of photodetectors.

imaging optical components are independently located within a free space, precise alignment of such components in the submicrometer range is required. Therefore, both mechanical and temperature stabilities are crucial for the realization of free-space optical interconnections. Of course, the fabrication cost of an optical interconnection system is also of practical interest.

To alleviate the difficulties of the free-space optical interconnection packaging, planar-optics-based interconnection schemes have been proposed.[4–10] As shown in Figure 5.2, the imaging path is folded within a compact planar-optics-based propagating medium. By employing this planar optics scheme, the advantages of precise alignment accuracy of the optical components by placing them on a single substrate and mechanical and thermal stability are expected.

However, there are obvious drawbacks associated with this planar optics approach. Because the beam propagation paths are folded, it is apparent that this planar optics approach does not fully utilize the potential of 3D free-space volume. The number of interconnection channels is limited because optical elements such as lenses and deflectors are located on the same side of the substrate and cannot physically overlap to function properly (see Figure 5.2). An oblique angle of propagation may introduce aberration into the propagating

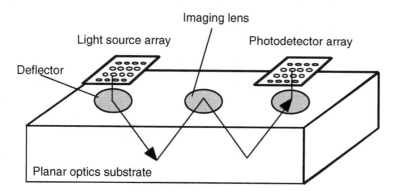

Figure 5.2 Schematic diagram of free-space optical interconnection based on planar optics substrate.

beam that will require careful optical system designs. In this chapter, various aspects of planar-optics-based free-space optical interconnections are reviewed and discussed in terms of optical system design, scalability, tolerancing, optical input and output interfaces, device packaging, and applications.

5.2 PLANAR OPTIC VARIATIONS

Various planar-optics-based systems have been proposed for implementing free-space optical interconnections. Beam propagation in a zigzag manner was originally proposed[4] by the combination of holograms and a mirror (Figure 5.3). Emitted light from the light source is collimated and deflected by the hologram. The light is reflected by the mirror and finally directed to the photodetector by the other hologram. A pair of holograms is required to establish a single optical link. Therefore, the resulting interconnection density will be relatively limited. In principle, multiple holograms are superimposed by recording the holograms with different angles or wavelengths.[11,12] However, adding such multiplexing capability may make the optical system complicated and impractical. Another drawback of using holograms for planar optical interconnection is difficulty in replication. Prior to use of the hologram, the hologram must be recorded by interfering mutually coherent light beams. An inexpensive replication method for volume holograms has not been established yet, making it impractical

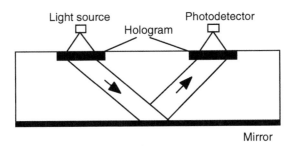

Figure 5.3 Folded free-space optical interconnection consisting of holograms and a reflecting mirror. (K.-H. Brenner and F. Sauer, *Appl. Opt.*, 27, 4251, 1988.)

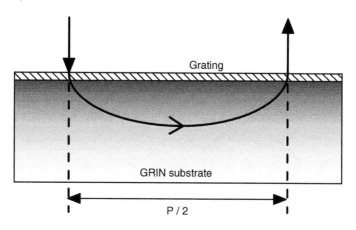

Figure 5.4 Planar optical interconnection based on the combination of a grating and a GRIN substrate.

to realize inexpensive optical interconnection based on volume holograms.

Optical interconnections based on gradient index (GRIN) planar optics have been proposed.[5,6] The proposed planar interconnection system is shown in Figure 5.4. Emitted light from a light source is introduced into the GRIN substrate by the top grating. Then, in the GRIN substrate, the incident light follows a curved trajectory and emerges normally from the GRIN substrate. The distance between the input and output locations on the grating is limited to a regular pitch $P/2$, which is equal to the half pitch of conventional GRIN lenses. It is also pointed out that it is difficult to make the optical axes of two GRIN rod lenses cross, making this approach limited with relatively regular interconnection topologies. Mass producibility of this approach is yet to be demonstrated.

Similar to the GRIN planar optics approach, making a planar optical interconnection medium based on a refractive lens has been proposed.[7] The proposed planar optics is schematically shown in Figure 5.5. Nearly ideal spherical geometry of index of refraction distribution was experimentally reported using a field-assisted ion exchange process.

Figure 5.6 shows one of the most promising approaches: a planar optical imaging system consisting of off-axis lenses.[8–10]

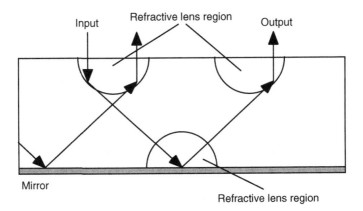

Figure 5.5 Free-space optical interconnection implemented using a planar refractive lens substrate. (J. Bahr and K.-H. Brenner, *Proc. SPIE*, 3490, 419, 1998.)

The lenses are either diffractive[13–15] or refractive types.[16,17] The incident light from a light source is collimated and deflected by the microlens array. To prevent the collimated beams from diverging because of diffraction, a pair of relay lenses is used. In this approach, all the necessary optical elements are fabricated on a single side of the propagating substrate. By using photolithography and etching techniques routinely used in

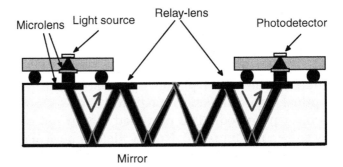

Figure 5.6 Free-space optical interconnection based on planar optics consisting of off-axis imaging lenses and a reflecting mirror.

mature Si large-scale integrated (LSI) industry, precise placements with submicrometer accuracy of the optical elements are easily accomplished. This wafer-scale process yields mass production at low cost.

To control both focal length and deflection angle rigorously, the diffractive optical element (DOE) is preferable. However, DOEs tend to result in lower diffraction efficiency than refractive lenses for larger deflection angles because the corresponding grating period becomes smaller. Parameters such as substrate thickness and deflection angle are closely coupled and directly affect the number of total interconnection channels implemented.[18] A larger deflection angle leads to a thinner propagating substrate for a given number of interconnection channels. For the constant substrate thickness, the larger deflection angle results in fewer reflections. Although total internal reflection can realize virtually no reflection losses,[19,20] corresponding critical angles of 41.8° in silica or 16.6° in Si are relatively large to be realized by DOEs.

To realize a large deflection angle without sacrificing insertion losses, lithographically fabricated microprisms have been reported.[21-23] Gimkiewicz et al.[23] reported a microprism with sags ranging from 6 to 20 μm with a base area of 100 μm² in photoresist. The transferred prism height in glass is approximately 4.2 μm. It is reported that prisms fabricated with a large index of refraction, such as GaAs and Si, are preferable because the corresponding deflection angle is much larger than low-index material for a given sag height. The fabrication accuracy of the microprism currently needs further process development for use for planar optical interconnection systems. It is also important to employ tolerant optical designs to cope with the possible fabrication errors of refractive lenses and prisms used.

5.3 OPTICAL SYSTEM CONSIDERATIONS

Because planar-optics-based interconnections are simply folded versions of free-space optical interconnections, optical system considerations aimed at free-space optical interconnections are still applicable.[24-26] Lohmann[25] divided the free-space

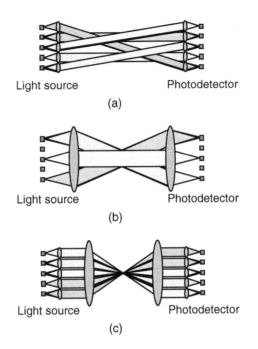

Light source   Photodetector
(a)

Light source   Photodetector
(b)

Light source   Photodetector
(c)

Figure 5.7 Schematic diagram of free-space optical interconnections: (a) micro-, (b) macro-, and (c) hybrid imaging systems. (A.W.
Lohmann, *Opt. Commun.*, 86, 365, 1991.)

optical interconnection systems into three categories: micro-,
macro-, and hybrid systems (Figure 5.7).

In the microsystem, a microlens is assigned for each light
source and a photodetector. Emitted light from the light
source is collimated and directed into an arbitrary direction
by the corresponding microlens. If the microlens pitch is
small, the propagating beam diverges because of diffraction.
The diverging beam couples into the adjacent microlens, causing cross talk. Therefore, the density of a microsystem is
relatively limited.

For the macroimaging system, an array of light sources
is imaged onto an array of photodetectors by either a single
lens (2-f imaging) or a pair of macrolenses (4-f imaging). For
the same number of array sizes and a diverging angle of light
sources, the aperture size of the macrolens is larger for a 2-f

imaging system than that for a 4-f imaging system. A well-corrected macroimaging lens is capable of imaging interconnection channels on the order of 10^6. However, this high level of resolution tends to be wasted if the light source is "dilute," that is, the light source pitch is much larger than the light source size, and the light sources are sparsely located in the light source array chip.[25] This is the case with the so-called smart pixel array,[27] in which the semiconductor die is partitioned into an array of subpixels with the light source, photodetector, and some functional electronic circuits. Because each light source needs a driving circuit and a photodetector needs an amplifier as well as logic circuits, the subpixel pitch is usually 100 to 250 µm.

The hybrid system is a combination of the micro- and macrosystems. The emitted light is collimated by the nearby microlens. Then, the bundle of collimated beams is imaged by a pair of macrolenses. In this hybrid imaging system, the imaging task is divided into the microlens and the macrolens. Because the emitted light is collimated, the aperture size required for the macrolens is smaller than that for the macrosystem. It should also be noted that the diffraction effect at the microlens is compensated for by the macroimaging.

In the actual system design, the choice of proper optical system depends on the interconnection length, the number of interconnection channels, the pitch of the channels, and tolerancing conditions of each optical element involved. Interconnection length L is given by Equation 5.1 for a predetermined value of beam spot size ω.[28]

$$L = \frac{\pi}{\lambda} \omega^2 \qquad (5.1)$$

The corresponding interconnection lengths L are plotted as a function of ω in Figure 5.8. For the calculation, the wavelength was assumed to be $\lambda = 1$ µm. When the interconnection length is short, a macroimaging system may work just as well as the hybrid system. If the channel pitch can be set at a relatively large value to ignore the diffraction at a microlens, even the microsystem can fulfill the interconnection requirement, making the required optical system simple.

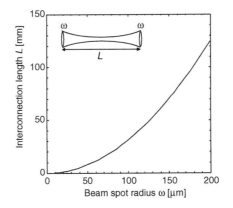

Figure 5.8   Interconnection length *L* as a function of beam spot radius ω determined by diffraction. (H. Sasaki, K. Shinozaki, and T. Kamijoh, *Opt. Eng.*, 35, 2240, 1996.)

5.4 SYSTEM SCALABILITY

In this section, interconnection system scalability is considered assuming the planar optics substrate based on DOEs (Figure 5.9).[18] The fabricated prototype optical substrate is shown in Figure 5.10. Collimated beams are assumed to be incident into the input linear grating, deflected into the optical substrate,

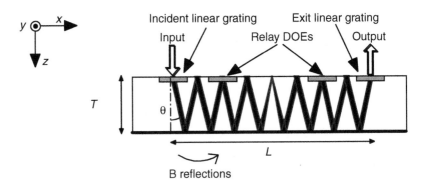

Figure 5.9   Schematic diagram of the optical system considered for scalability analysis. (H. Sasaki, K. Kotani, H. Wada, T. Takamori, and T. Ushikubo, *Appl. Opt.*, 40, 1843, 2001.)

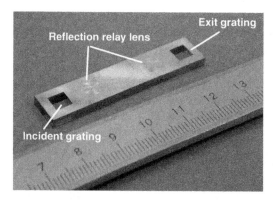

Figure 5.10 Photograph of the fabricated planar-optics-based free-space interconnection system. (H. Sasaki, K. Kotani, H. Wada, T. Takamori, and T. Ushikubo, *Appl. Opt.*, 40, 1843, 2001.)

and finally to exit through the output linear grating. The beam waist of the incident collimated beam is assumed to be at the incident linear grating. Two reflection-type relay DOEs are fabricated on the top surface of the substrate. The incident light is reflected B times before reaching the first relay lens. The interconnection length L is defined as the lateral distance between the input and the output beam locations. The thickness of the substrate is set as T. The deflection angle of the linear grating is θ. Both surfaces of the glass substrate, except for input and output gratings, were mirror coated. Figure 5.11 indicates the multilevel binary structure of DOEs.[13–15] The smallest step width of the DOE is called the *minimum feature size*. In general, DOEs are fabricated by repeating multiple

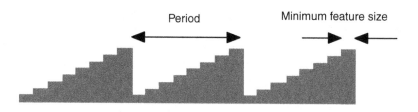

Figure 5.11 Multilevel binary structure of the DOE.

photolithography and etching processes. Because extremely small minimum feature size leads to fabrication difficulty, the minimum feature size of the DOEs determines the manufacturability of the target optical system.[29]

In practice, DOEs with extremely small minimum feature sizes yield to lower diffraction efficiency than the value predicted by scalar diffraction theory. However, it is currently impractical to explicitly take into account the effect of such diffraction efficiency degradation in a planar-optics-based system because vector diffraction theory[30] needs to be solved numerically at an entire aperture of each DOE. Therefore, the entire portion of the following theoretical derivation assumes that the designed DOEs provide diffraction efficiency large enough to make the optical interconnection system insertion loss acceptable.

Figure 5.12 summarizes the design flow of the photonic circuit discussed in this section. First, the interconnection

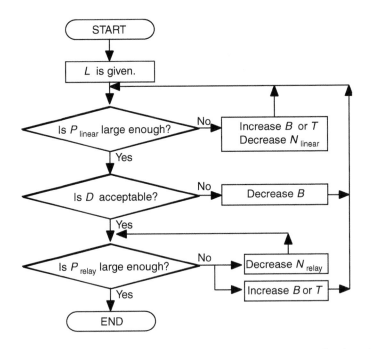

Figure 5.12 Planar optical interconnection system design flow based on a DOE.

length L is given. Then, the minimum feature size of a pair of linear gratings, P_{linear}, is calculated. If the minimum feature size is too small to be implemented, there are three choices: increase the substrate thickness T, increase the number of beam reflections B, or decrease the number of phase levels of the grating N_{linear}. The second step is to examine the maximum number of interconnection channels D in the x–z plane in Figure 5.9. If we need more interconnection channels, the only remaining approach is to decrease the number of beam reflections B because increasing the substrate size does not help much. Note that decreasing the number of beam reflections B also makes the minimum feature sizes of the linear grating smaller since the deflection angle becomes larger. Therefore, either the minimum feature size of the grating or the maximum number of interconnection channels must be sacrificed to continue the design process. The third step is to examine the minimum feature sizes of the relay lenses P_{relay}. This requirement is usually relaxed. If there is a confrontation, there are two choices at this stage. The first choice is to decrease the phase level of the relay lens N_{relay}. This has little effect on the previous design process. The second choice is to increase either the number of beam reflections B or the thickness of the substrate T. As shown in Figure 5.9, increasing the number of beam reflections leads to fewer maximum interconnection channels D. If the maximum number of interconnection channels must be kept, the substrate thickness T must be increased.

Figure 5.13 shows the calculated minimum feature size of the incident linear grating as a function of the substrate thickness T and the interconnection length L. The following values are assumed for the calculations: The number of reflections $B = 2$, the number of phase levels of the linear grating $N_{linear} = 2$, the index of refraction of the silica optical substrate $n = 1.444$, and the wavelength of the light $\lambda = 1.55$ μm. For a constant value of the substrate thickness, increasing the interconnection length results in a smaller minimum feature size for the linear grating. For a given value of the interconnection length, increasing the substrate thickness alleviates the minimum feature size requirement for the linear grating.

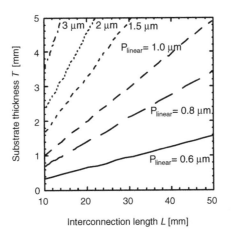

Figure 5.13 Minimum feature size of the binary linear gratings for incident beam deflection.

In addition, larger reflection B results in a smaller deflection angle θ, leading to larger minimum feature sizes. On the other hand, the increased phase levels N_{linear} of course make the corresponding linear grating's minimum feature sizes P_{linear} smaller. After evaluating the values in Figure 5.13, we chose an interconnection length L = 40 mm and a substrate thickness T = 3 mm. The resulting minimum feature size of the linear grating was then calculated as 0.84 μm.

As shown in Figure 5.9, optical beams are reflected and propagate in the x–z plane. Because the plane at which the beams are reflected on the top surface of the substrate must be physically separated from the relay lens to realize zigzag beam propagation, this physical separation leads to a scalability limitation. The maximum number of light beams D in the x–z plane (see Figure 5.14) has been calculated and is shown as a function of the incident beam waist radius ω in Figure 5.15. The following parameters were used in the calculations: B = 2, and L = 40 mm. The corresponding total width of channels along the x-axis determined by $4D\omega$, where ω is the incident beam waist radius at the linear grating, is also shown in the same figure. The number of channels multiplexed

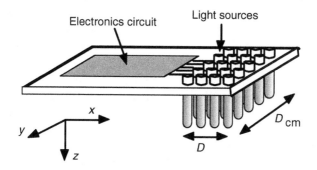

Figure 5.14 Schematic diagram describing the number of free-space optical interconnections along the x-axis.

in the x–z plane becomes larger for smaller values of the incident beam waist ω.

It is reasonable to assume that light sources can be integrated along the y-axis at the same density as along the x-axis, as shown in Figure 5.14. Here, the linear channel density D_{cm} is defined as a total number of free-space optical

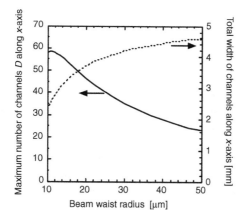

Figure 5.15 Maximum number of channels D and the total width of channels of the free-space optical interconnections along the x-axis as a function of the incident beam waist radius.

interconnection channels per 1 cm along the periphery of the optoelectronic chip (see Figure 5.14). The calculated results are shown in Figure 5.16 for $B = 2$ and $L = 40$ mm. Interconnection densities exceeding 1000 are available for a wide range of beam waist radii ω, demonstrating the possibility of the planar-optics-based free-space optical interconnections. Although the input light sources are initially assumed to be densely packaged as shown in Figure 5.14, the results in Figure 5.16 imply that relatively dense interconnections are feasible for smart-pixel-type light sources. For example, the incident beam waist of 50 µm still gives more than 1000 interconnections. The beam waist of 50 µm corresponds to the light source pitch of 200 µm, suitable as a pixel pitch in the smart pixels.

At the end of this section, I briefly mention the manufacturability of the two reflection-type relay DOEs. For the interconnection length $L = 40$ mm, the substrate thickness $T = 3$ mm, and the eight phase levels $N_{relay} = 8$, the corresponding minimum feature size of the relay DOEs is around 1 µm, which is readily fabricated by the current level of photolithography and etching techniques.

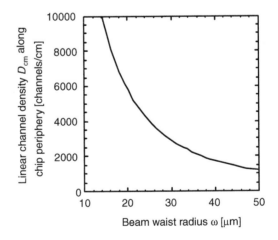

Figure 5.16 Interconnection density D_{cm} along the chip periphery of the free-space optical interconnections as a function of the incident beam waist radius ω.

5.5 OPTICAL SYSTEM TOLERANCING ANALYSIS

To mass-produce planar-optics-based optical interconnections at low cost, it is crucial to design the optical system with enough tolerance against fabrication errors and operating condition variations. Various research studies have reported on optical system tolerancing issues.[18,31–36] Tolerancing conditions considered are substrate expansion caused by temperature variation, wavelength used, substrate thickness, parallelism (wedge angle) of the substrate, and light source and photodetector mechanical registration errors. Such tolerancing condition fluctuations make the output beam position laterally shift and cause coupling efficiency degradation and interchannel cross talk after propagating in a zigzag manner inside the planar optics substrate. Among such tolerancing conditions, thermal effect is usually negligible.[31] It was found that substrate parallelism has the worst effect on the lateral dislocation of the output beam. The lateral shift of the beam output position shifts exponentially as the wedge angle of the substrate increases.

Lunitz and Jahns[36] proposed to construct a planar-optics-based optical interconnection using a series of lenses in which each lens is separated by the focal length of the lens used (Figure 5.17a). This system is optically equivalent to a conventional confocal resonator (Figure 5.17b). In the confocal resonator, the laterally offset beam is pulled back to its correct location, and the effect of misalignment is periodically reset, leading to a stable system. A series of lenses works the same way and has been well known as a "light pipe" or "lens waveguide." By introducing such optical configuration into a planar optical system, the resulting optical interconnection is expected to be stable against many tolerancing factors.

Sasaki et al.[18] compared tolerancing characteristics of three different optical system configurations with both interconnection length and substrate thickness fixed. The three different configurations are compared in Figure 5.18. Type I is an identical configuration to that shown in Figure 5.9. Types I and III have the same angle of beam propagation, $\theta = 39.8°$. In contrast, Type II has twice as many reflections as Types I and II, leading to a smaller angle of propagation, $\theta = 19.9°$.

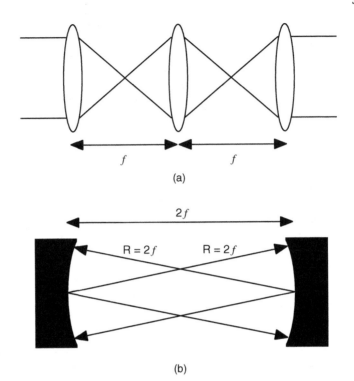

Figure 5.17 (a) Lens waveguide system and (b) corresponding confocal resonator with a pair of convex mirrors to yield to tolerant free-space optical interconnections. (B. Lunitz and J. Jahns, *Opt. Commun.*, 134, 281, 1997.)

Type III has two sets of cascaded 4-f imaging lenses compared with Type I. Note that both decreasing the propagation angle by half as in Type II and cascading relay lenses as in Type III decrease the total number of interconnections because of the physical separation of the propagating beams and DOEs. Interconnection length is set at 40 mm, and the thickness of the optical substrate is 3 mm.

Similar tolerance analysis of planar optical interconnection was reported by Zaleta et. al.,[34] who numerically analyzed the tolerance of a microimaging configuration with relay lenses in terms of the optical power vignetting at each DOE. Unlike the throughput power loss in a microimaging system, an optical

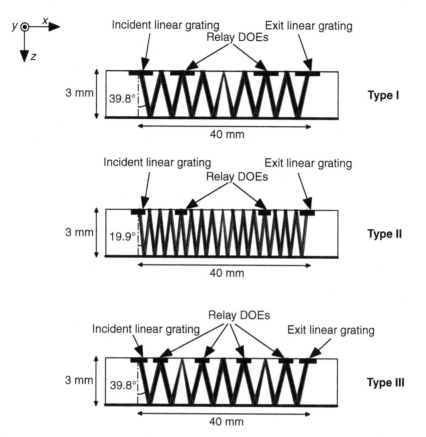

Figure 5.18 Three different planar-optics-based interconnection configurations. (H. Sasaki, K. Kotani, H. Wada, T. Takamori, and T. Ushikubo, *Appl. Opt.*, 40, 1843, 2001.)

beam coupling into adjacent interconnection channels and the corresponding signal-to-noise ratio degradation may have severe tolerance implications on tolerancing in the hybrid imaging system, even with modest levels of incident power at the detector. To take into account the effect of off-axis beam propagation, the collimated light beam was assumed to be incident on the incident linear grating surface but offset by 450 μm along both x- and y-axes. After the ray trace simulations performed using an optical computer-aided design (CAD)

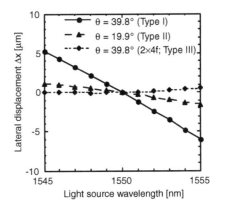

Figure 5.19 Optical system tolerances in terms of wavelength of the propagating beam.

program, the lateral position shift of the exiting light beam at the exit linear grating was used as a measure of the optical system tolerance. Because the lateral shift along the x-axis is always greater than that along the y-axis for the three given configurations, only the lateral shift along the x-axis was shown.

Because the DOE's focal length and deflection angle have a strong wavelength dependence, different wavelengths other than the designed value produce beam spot wandering at the exit linear grating (Figure 5.19). The propagating beam wavelength was assumed to have a temperature-dependent variation of $\Delta\lambda = \pm 5$ nm. As shown in Figure 5.19, reducing the propagation angle (Type II) results in better wavelength tolerance compared to Type I. However, maintaining the relatively large propagation angle with a cascaded relay lens system (Type III) resulted in even better wavelength tolerance than with Type II.

The optical substrate thickness tolerance is shown in Figure 5.20. Although optical substrates with better than 2″ parallelism are readily available, the absolute value of the substrate thickness tends to vary by ± 100 μm. Reducing the propagation angle yields better tolerance for thinner substrates than the designed thickness; this improvement is less noticeable for thicker substrates compared to the designed value. In contrast, Type III shows a wide substrate thickness tolerance.

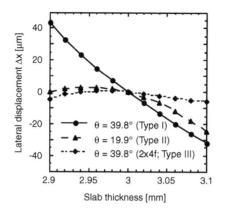

Figure 5.20 Optical system tolerances in terms of optical system substrate thickness.

Figure 5.21 shows simulation results on the substrate parallelism (wedge angle) tolerance for the three types of photonic circuits. Optical substrates with parallelism that is in the range of 2″ to 5″ are readily available for surface areas larger than 10 by 10 cm. In the calculations, the bottom surface of the propagating substrate was rotated by a specified number of degrees about both the x- and y-axes. It should be noted that, although reducing the propagation angle (Type II)

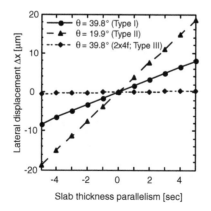

Figure 5.21 Optical system tolerances in terms of the parallelism (wedge angle) of the optical substrate.

produced better tolerance in terms of wavelength and substrate thickness variations than Type I, substrate parallelism affected the Type II configuration more severely than Type I. This may be explained by the fact that the smaller angle of propagation leads to more reflections, and the effect of imperfect parallelism affects the propagating beam at every reflection. In contrast, Type III showed good tolerance for the given conditions.

Comparing the above tolerance analyses for the three different photonic circuit configurations, cascading relay lenses seems to be the most promising approach. Although decreasing the propagation beam's deflection angle (Type II) showed good tolerance against wavelength and substrate thickness deviations, it is found to be sensitive to substrate parallelism. In contrast, the cascaded relay lens approach (Type III) demonstrated robustness for all three tolerance conditions considered and exhibited an exit beam wandering of $\Delta x < 7$ μm for all given parameters. This level of lateral beam shift can be completely compensated for by assuming an ample margin for the microlens pitch.

5.6   OPTICAL INTERFACES

5.6.1   Light Sources

To realize the compact packaging of planar-optics-based free-space optical interconnections, a densely packaged light source array is required. The light source can be either a modulator array[27] or a vertical cavity surface-emitting laser (VCSEL).[37–41] To make the optical system simple, a VCSEL array is preferred because modulators need an additional light source and corresponding optical system to direct the light to the modulator.

As mentioned, emitted light from the light source is collimated by a nearby microlens in either a micro- or hybrid imaging system. However, aligning a microlens array and a light source array with high precision is a complicated and time-consuming task. To solve this alignment difficulty, the monolithic integration of a semiconductor microlens array and a VCSEL array was proposed.[42,43] An array of VCSELs with AlGaAs/GaAs mirrors and an active region with three InGaAs

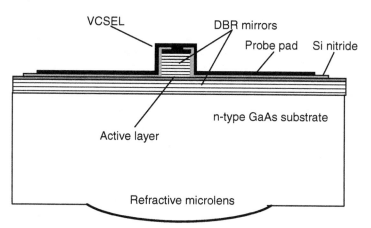

Figure 5.22 Schematic diagram of a VCSEL integrated with a refractive microlens on the back surface of the substrate. (E.M. Strzelecka, G.D. Robinson, M.G. Peters, F.H. Peters, and L.A. Coldren, *Electron. Lett.*, 31, 724, 1995.)

quantum wells was grown by molecular beam epitaxy on the n-type GaAs substrate (see Figure 5.22). After the VCSEL fabrication, a deep ultraviolet (UV) photoresist was circularly patterned on the rear surface by aligning the VCSEL and the photoresist centers with the aid of an infrared mask aligner. The photoresist was reflowed in an oven, and a near-parabolic shape of reflowed photoresist was obtained. This photoresist shape was transferred into the GaAs substrate by a reactive ion etching process. The monolithic integration of a photodetector and a refractive lens based on a similar fabrication process was also reported.[43]

In smart pixels, logical functions are realized in Si complementary metal-oxide semiconductor (CMOS) circuits. Therefore, it is even favorable to integrate Si CMOS circuits with a combination of a light source and the corresponding microlens. However, optical devices such as lasers and photodetectors are based on III–V semiconductor materials. Therefore, reliable integration between different types of semiconductors is required.

Wada et al.[44] reported a wafer-bonding technology between Si and InP substrates aiming at the integration of a VCSEL

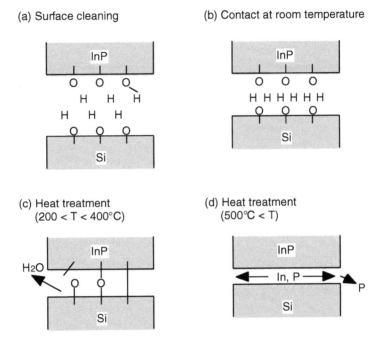

Figure 5.23 Direct bonding process mechanism. (H. Wada, H. Sasaki, and T. Kamijoh, *Solid-State Electron.*, 43, 1655, 1999.)

and a microlens on Si substrate at wafer scale. The bonding mechanism is illustrated in Figure 5.23. First, OH groups are absorbed on the wafer surface during the cleaning process with H_2SO_4:H_2O_2:H_2O (see Figure 5.23a). When the cleaned surfaces are brought into contact at room temperature, the wafers adhere to each other through the hydrogen bonding between the OH groups (see Figure 5.23b). If the temperature is elevated to moderate values between 200 and 400°C, water molecules start to evaporate and escape from the interface, and the hydrogen bonds are replaced by bonds such as InP-O-Si (Figure 5.23c). This bond is usually stronger than the hydrogen bond, which results in increased strength at higher temperatures. When the wafers are annealed at higher temperatures ($T > 500$°C), evaporation of phosphorus from InP and the migration of In atoms become pronounced, which can cause atomic rearrangement at the interface (see Figure 23d).

(a) Preparation of InP DH film

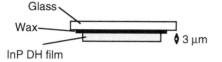

(b) Island formation

(c) Surface cleaning and wafer bonding

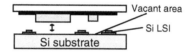

(d) Removal of glass plate and annealing

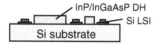

(e) Optical device fabrication

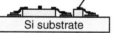

(f) Back surface DOE fabrication

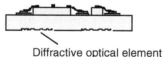

Diffractive optical element

Figure 5.24 Schematic diagram of the integration of optical devices on a Si LSI wafer. (H. Wada, H. Sasaki, and T. Kamijoh, *Solid-State Electron.*, 43, 1655, 1999.)

Then, the two wafers can be atomically bonded to form stronger bonds.

The integration of optical devices onto Si LSI substrates is illustrated in Figure 5.24. First, islands are formed on the InP double-heterostructure (DH) thin film by photolithography

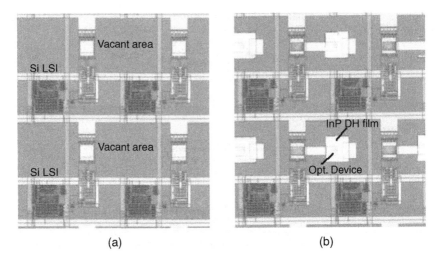

(a)   (b)

Figure 5.25 (a) Vacant area on Si LSI array. (b) Optical device was fabricated on the vacant area and connected to the Si LSI circuit.

and etching corresponding to the vacant space on the Si LSI substrate (see Figure 5.24a and Figure 5.24b). The surfaces of both wafers are cleaned, and the islands are aligned and directly bonded at room temperature (Figure 5.24c). After removing the glass plate and annealing at 400°C (Figure 5.24d), optical devices are fabricated on the DH islands by lithography, and the interconnection metals are deposited and patterned (Figure 5.24e). A DOE working as a backsurface collimator was finally fabricated on the rear side of the Si substrate (Figure 5.24f). Figure 5.25a shows an array of Si LSI circuits with vacant areas. Optical devices are fabricated on the vacant area as shown in Figure 5.25b.

5.6.2 Optical Fibers

In the planar-optics-based free-space optical interconnections, propagated optical signals may not necessarily be terminated at photodetectors. There may be a need to couple the optical signals into a bundle of optical fibers for further transmission to other planar-optics-based modules.

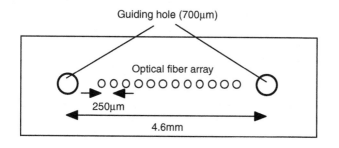

Figure 5.26 Cross-section view of 12-channel MT connector.

An array of optical fiber interfaces have been recently reported based on MT connectors.[45–47] In the conventional MT connectors, arrays of 2, 4, 8, 10, and 12 optical fibers are available with a 250-μm pitch. Figure 5.26 shows a cross section of the MT connector. Optical fibers are either single mode or multimode. To align the MT connector, a pair of 0.7-mm diameter guiding holes is located at both sides of the connector. These two guiding holes are separated by 4.6 mm.

Gruber et al.[47] fabricated alignment rods by milling them from the bulk aluminum alloy substrate using a computer numerical control machine. Then, the MT connector was precisely aligned by inserting these alignment rods into the guiding holes of the connector. Optical coupling between two MT connectors was successfully demonstrated using planar optics in conjunction with a metal MT connector interface (Figure 5.27). Multimode fibers were used. Note that conventional multimode fibers are capable of transmitting 1 Gb/sec optical signals more than 100 m. Therefore, multimode fibers are more than enough for short-distance optical interconnections. Coupling efficiency variation over 10 channels was less than 10%.

5.6.3 Optical Device Packaging

Precise alignment between the planar optics substrate and an array of light sources or photodetectors is crucial for free-space optical interconnections. For the microimaging system, required alignment tolerance between the light source and the

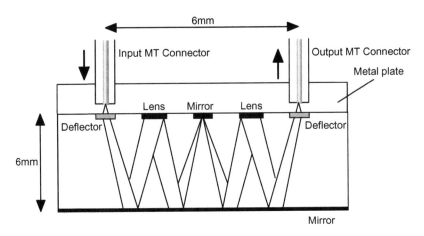

Figure 5.27 MT-connector-to-MT-connector interconnection based on planar optics. A pair of MT connectors is aligned and fixed on the metal plate placed on the planar optics substrate. (M. Gruber, J. Jahns, E.M.E. Joudi, and S. Sinzinger, *Appl. Opt.*, 40, 2902, 2001.)

nearby microlens is stringent, typically on the order of micrometers. Misalignment between the light source and the microlens causes the beam deflection angle error and leads to cross-talk degradation. Although the placement accuracy of the light source is somewhat relaxed for a macroimaging system, a photodetector must be precisely aligned against the light source image to detect the incoming optical signal efficiently.

When the microlenses are separately fabricated on the surface of the planar optics substrate, this stringent alignment between the light source array and the microlens array must be established, making the packaging process difficult. As mentioned, wafer-scale integration of light sources and corresponding microlenses is promising for solving this difficulty.[42–44] Once the emitted light from the light source is collimated, the alignment accuracy requirement of such a bundle of collimated beams is relaxed (i.e., around 10 μm) compared with the microlens alignment accuracy.

For the packaging of optical devices on the planar optics substrate, various packaging technologies have been adopted.[48–54]

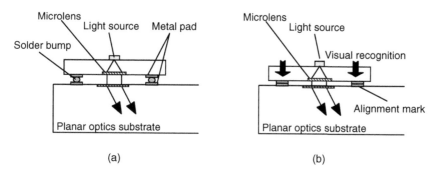

(a)   (b)

Figure 5.28 Packaging of optical device onto the planar optics substrate by (a) solder bump and (b) visual recognition of alignment marks.

Figure 5.28a and Figure 5.28b show the optical device packaging based on solder bump[49-51] and visual alignment[52-54] technologies. Each optical device may contain hundreds of light sources. Regardless of the number of optical channels, all the necessary alignments can be accomplished with a one-time alignment process.

For the solder bump alignment process, metal pads are prepared on both surfaces of the planar optics substrate and the optical device. Then, the solder alloy is placed on these metal pads during wafer fabrication. The solder is heated above the melting point so that solder balls are formed. The optical device is placed on the contact with sufficient force. Finally, the environmental temperature is again heated above the solder melting point so that the solder reflows. The surface tension of the solder under the liquid state makes the device self-align to the planar optics substrate. Lateral misalignments of ± 2.0 μm[50] and ± 0.5 μm[51] were reported. In contrast to the lateral alignment accuracy, vertical alignment variation of the solder bump is usually larger unless the volume of each solder bump is precisely controlled. However, the lack of alignment accuracy along the vertical direction is not problematic for the packaging of a VCSEL array on the planar optics substrate.

The alignment process by visually recognizing the alignment marks on the optical device and the planar optics substrate as shown in Figure 5.28b is another promising technology. Such alignment marks are easily and precisely deposited on both surfaces by photolithography at wafer scale. These alignment marks are visually recognized by an infrared charge-coupled device (CCD) camera. By visually aligning corresponding marks, lateral alignment accuracy of ±1.0 μm have been demonstrated.[52–54]

5.7 APPLICATIONS

The applications of planar-optics-based free-space optical interconnections are not limited to one-to-one interconnections. For example, space-variant operation such as fan out can be easily implemented by inserting a computer-generated hologram at a Fourier transform plane in the optical system.[55]

Song et al.[56] implemented perfect shuffle interconnections based on planar optics. Optical correlators[57–59] and packet address detection for wavelength division multiplexing (WDM) optical signals[60] have been proposed in a planar-optics-based system. Other applications demonstrated include multichannel Fourier transform,[61] a vector–matrix multiplier,[62] and multistage interconnection networks.[63] In this section, I review the planar-optics-based implementation[63] of multistage interconnection networks.[64]

The original multistage interconnection network has five stages. A Banyan interconnection is used to implement an extended generalized shuffle network. The detail of the optical system is depicted in Figure 5.29a for a single stage. The switching function is realized by the reflection-type smart pixel based on the field effect transistor–self-electro-optic effect device (FET–SEED) technology.[65] Note that only one input signal is traced in Figure 5.29a and Figure 5.29b for the sake of simplicity. First, the input signal from the previous stage is detected. Based on this input signal, the signal from the light source is modulated by the smart pixel and transferred to the next stage. A quarter-wave plate is used to rotate the polarization of the modulated signal to direct it to the next stage. A one-by-three fan-out device based on a computer-generated hologram is used to implement Banyan interconnection. The entire five-stage interconnection was demonstrated mounted

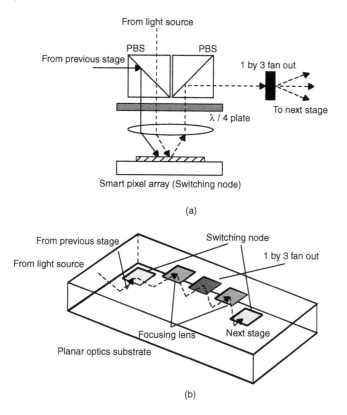

Figure 5.29 Multistage interconnection based on (a) optome-chanical configuration (F.B. McCormick, T.J. Cloonan, A.L. Lentine, J.M. Sasian, R.L. Morrison, M.G. Beckman, S.L. Walker, M.J. Wojcik, S.J. Hinterlong, R.J. Crisci, R.A. Novotny, and H.S. Hinton, *Appl. Opt.*, 33, 1601, 1994); and (b) planar optics (S.H. Song, J.-S. Jeong, and E.-H. Lee, *Appl. Opt.*, 36, 5728, 1997).

on a 28- by 35-cm metal mounting plate. Lateral alignment tolerance was ±2 µm, making packaging of the system extremely difficult.

Song et al.[63] proposed a planar-optics-based implementation of a multistage interconnection similar to that shown in Figure 5.29b. A smart pixel array is placed on the top surface of the optical substrate. The optical signal from the previous stage is propagated in a zigzag manner and detected by the smart pixel array. This smart pixel array modulates the other incoming optical beam from the light source, and the modulated signal further propagates within the optical substrate and finally reaches the next switching node through imaging optics. By integrating necessary optical elements on the top surface, precise alignment among those elements is expected. Note also that a planar-optics-based system can greatly reduce the number of alignments during packaging, making the system more reliable. It is also pointed out that the resulting optical system size may be greatly reduced in volume compared with the original version.

REFERENCES

1. J.W. Goodman, F.J. Leonberger, S.-Y. Kung, and R.A. Athale, Optical interconnections for VLSI systems, *Proc. IEEE*, 72, 850 (1984).

2. M.R. Feldman and C.C. Guest, Interconnect density capabilities of computer generated holograms for optical interconnection of very large scale integrated circuits, *Appl. Opt.*, 28, 3134 (1989).

3. L.J. Camp, R. Sharma, and M.R. Feldman, Guided-wave and free-space optical interconnects for parallel-processing systems: a comparison, *Appl. Opt.*, 33, 6168 (1994).

4. K.-H. Brenner and F. Sauer, Diffractive-reflective optical interconnects, *Appl. Opt.*, 27, 4251 (1988).

5. S.H. Song, S. Park, C.H. Oh, P.S. Kim, M.H. Cho, and Y.S. Kim, Gradient-index planar optics for optical interconnections, *Opt. Lett.*, 23, 1025 (1998).

6. E.-H. Lee and S.H. Song, Three-dimensional planar-integrated optics: a comparative view with free-space optics, *Proc. SPIE*, 3950, 194 (2000).

7. J. Bahr and K.-H. Brenner, Optical motherboard: a planar chip to chip interconnection scheme for dense optical wiring, *Proc. SPIE*, 3490, 419 (1998).

8. J. Jahns and A. Huang, Planar integration of free-space optical components, *Appl. Opt.*, 28, 1602 (1989).

9. J. Jahns and B. Acklin, Integrated planar optical imaging system with high interconnection density, *Opt. Lett.*, 18, 1594 (1993).

10. J. Jahns, Planar packaging of free-space optical interconnections, *Proc. IEEE*, 82, 1623 (1994).

11. J.M. Heaton, P.A. Mills, E.G.S. Page, L. Solymar, and T. Wilson, Diffraction efficiency and angular selectivity of volume phase holograms recorded in photorefractive materials, *Optica Acta*, 31, 885 (1984).

12. F.H. Mok, M.C. Tackitt, and H.M. Stoll, Storage of 500 high-resolution holograms in a LiNbO$_3$ crystal, *Opt. Lett.*, 16, 605 (1991).

13. G.J. Swanson and W.B. Veldkamp, Diffractive optical elements for use in infrared systems, *Opt. Eng.*, 28, 605 (1989).

14. G.J. Swanson, Binary optics technology: the theory and design of multi-level diffractive optical elements, *MIT Tech. Rep.*, 854 (1989).

15. G.J. Swanson, Binary optics technology: theoretical limits on the diffraction efficiency of multilevel diffractive optical elements, *MIT Tech. Rep.*, 914 (1991).

16. W. Däschner, P. Long, R. Stein, C. Wu, and S.H. Lee, General aspheric refractive micro-optics fabricated by optical lithography using a high energy beam sensitive glass gray-level mask, *J. Vac. Sci. Technol. B*, 14, 3730 (1996).

17. W. Däschner, P. Long, R. Stein, C. Wu, and S.H. Lee, Cost-effective mass fabrication of multilevel diffractive optical elements by use of a single optical elements by use of a single optical exposure with a gray-scale mask on high-energy beam-sensitive glass, *Appl. Opt.*, 36, 4675 (1997).

18. H. Sasaki, K. Kotani, H. Wada, T. Takamori, and T. Ushikubo, Scalability analysis of diffractive optical element-based free-space photonic circuits for interoptoelectronic chip interconnections, *Appl. Opt.*, 40, 1843 (2001).

19. J.-H. Yeh and R.K. Kostuk, Free-space holographic optical interconnects for board-to-board and chip-to-chip interconnections, *Opt. Lett.*, 21, 1274 (1996).

20. R.T. Chen, F. Li, M. Dubinovsky, and O. Ershov, Si-based surface-relief polygonal gratings for one-to-many wafer scale optical clock signal distribution, *IEEE Photon. Technol. Lett.*, 8, 1038 (1996).

21. K. Reimer, U. Hofmann, M. Jurss, W. Pilz, H.J. Quenzer, and B. Wagner, Fabrication of microrelief surfaces using a one-step lithography process, *Proc. SPIE*, 3226, 2 (1997).

22. J. Jahns, C. Gimkiewicz, and S. Sinzinger, Light efficient parallel interconnect using integrated planar free-space optics and vertical cavity surface emitting laser diodes, *Proc. SPIE*, 3286, 52 (1998).

23. C. Gimkiewicz, D. Hagedorn, J. Jahns, E.-B. Kley, and F. Thoma, Fabrication of microprisms for planar optical interconnections by use of analog gray-scale lithography with high-energy-beam-sensitive glass, *Appl. Opt.*, 38, 2986 (1999).

24. J. Jahns and S.J. Walker, Imaging with planar optical systems, *Opt. Commun.*, 76, 313 (1990).

25. A.W. Lohmann, Image formation of dilute arrays for optical information processing, *Opt. Commun.*, 86, 365 (1991).

26. J. Jahns, F. Sauer, B. Tell, K.F. Brown-Goebeler, A.Y. Feldblum, C.R. Nijander, and W.P. Townsend, Parallel optical interconnections using surface-emitting microlasers and a hybrid imaging system, *Opt. Commun.*, 109, 328 (1994).

27. A.V. Krishnamoorthy and K.W. Goossen, Progress in optoelectronic-VLSI smart pixel technology based on GaAs/AlGaAs MQW modulators, *Int. J. Optoelectron.*, 11, 181 (1997).

28. H. Sasaki, K. Shinozaki, and T. Kamijoh, Reduced alignment accuracy requirement using focused Gaussian beams for free-space optical interconnection, *Opt. Eng.*, 35, 2240 (1996).

29. H. Sasaki, I. Fukuzaki, Y. Katsuki, and T. Kamijoh, Design considerations of stacked multilayers of diffractive optical elements for optical network units in optical subscriber-network applications, *Appl. Opt.*, 37, 3735 (1998).

30. M.G. Moharam and T.K. Gaylord, Diffraction analysis of dielectric surface-relief grating, *J. Opt. Soc. Am.*, 72, 1383 (1982).

31. J. Jahns, Y.H. Lee, C.A. Burrus, Jr., and J.L. Jewell, Optical interconnects using top-surface-emitting microlasers and planar optics, *Appl. Opt.*, 31, 592 (1992).

32. S.K. Patra, J. Ma, V.H. Ozguz, and S.H. Lee, Alignment issues in packaging for free-space optical interconnects, *Opt. Eng.*, 33, 1561 (1994).

33. F. Sauer, J. Jahns, C.R. Nijander, A.Y. Feldblum, and W.P. Townsend, Refractive-diffractive micro-optics for permutation interconnects, *Opt. Eng.*, 33, 1550 (1994).

34. D. Zaleta, S. Patra, V. Ozguz, J. Ma, and S.H. Lee, Tolerancing of board-level-free-space optical interconnects, *Appl. Opt.*, 35, 1317 (1996).

35. S. Sinzinger and J. Jahns, Integrated micro-optical imaging system with a high interconnection capacity fabricated in planar optics, *Appl. Opt.*, 36, 4729 (1997).

36. B. Lunitz and J. Jahns, Tolerant design of a planar-optical clock distribution system, *Opt. Commun.*, 134, 281 (1997).

37. J.L. Jewell, J.P. Harbison, A. Scherer, Y.H. Lee, and L.T. Florez, Vertical-cavity surface-emitting lasers: design, growth, fabrication, characterization, *IEEE J. Quantum Electron.*, 27, 1332 (1991).

38. D.I. Babi'c and J.J. Dudley, Double-fused 1.52-mm vertical-cavity lasers, *Appl. Phys. Lett.*, 66, 1030 (1995).

39. D.I. Babi'c, K. Streubel, R.P. Mirin, N.M. Margalit, J.E. Bowers, E.L. Hu, D.E. Mars, L. Yang, and K. Carey, Room-temperature continuous-wave operation of 1.54-mm vertical-cavity lasers, *IEEE Photonics Tech. Lett.*, 7, 1225 (1995).

40. T. Anan, M. Yamada, K. Nishi, K. Kurihara, K. Tokutome, A. Kamei, and S. Sugou, Continuous-wave operation of 1.30 mm GaAsSb/GaAs VCSELs, *Electron. Lett.*, 37, 566 (2001).

41. R. Shau, M. Ortsiefer, J. Rosskopf, G. Bohm, F. Kohler, and M.-C. Amann, Vertical-cavity surface-emitting laser diodes at 1.55 mm with large output power and high operation temperature, *Electron. Lett.*, 37, 1295 (2001).

42. E.M. Strzelecka, G.D. Robinson, M.G. Peters, F.H. Peters, and L.A. Coldren, Monolithic integration of vertical-cavity laser diodes with refractive GaAs microlenses, *Electron. Lett.*, 31, 724 (1995).

43. E.M. Strzelecka, G.B. Thompson, G.D. Robinson, M.G. Peters, B.J. Thibeault, M. Mondry, V. Jayaraman, F.H. Peters, and L.A. Coldren, Monolithic integration of refractive lenses with vertical cavity lasers and detectors for optical interconnections, *Proc. SPIE*, 2691, 43 (1996).

44. H. Wada, H. Sasaki, and T. Kamijoh, Wafer bonding technology for optoelectronic integrated devices, *Solid-State Electron.*, 43, 1655 (1999).

45. S. Sinzinger, Planar optics as the technological platform for optical interconnects, *Proc. SPIE*, 3490, 40 (1998).

46. S. Sinzinger, Systems engineering for planar-integrated free-space optics, *Proc. SPIE*, 4455, 256 (2001).

47. M. Gruber, J. Jahns, E.M.E. Joudi, and S. Sinzinger, Practical realization of massively parallel fiber-free-space optical interconnects, *Appl. Opt.*, 40, 2902 (2001).

48. V.N. Morozov, Y.-C. Lee, J.A. Neff, D. O'Brien, T.S. McLaren, and H. Zhou, Tolerance analysis for three-dimensional optoelectronic systems packaging, *Opt. Eng.*, 35, 2034 (1996).

49. J. Jahns, R.A. Morgan, H.N. Nguyen, J.A. Walker, S.J. Walker, and Y.M. Wong, Hybrid integration of surface-emitting micro-laser chip and planar optics substrate for interconnection applications, *IEEE Photon. Technol. Lett.*, 4, 1369 (1992).

50. B. Acklin and J. Jahns, Packaging considerations for planar optical interconnection systems, *Appl. Opt.*, 33, 1391 (1994).

51. W. Rehm, K. Adam, A. Goth, W. Jorg, J. Lauckner, J. Scherb, P. Aribaud, C. Artigue, C. Duchemin, B. Fernier, E. Grad, D. Keller, S. Kerboeuf, S. Rabaron, J.M. Rainsant, D. Tregoat, J.L. Nicque, A. Tournereau, P.J. Laroulandie, and P. Berthier, Low-cost laser modules for SMT, in *Technical Digest of Electronics Components and Technology Conference* (2000).

52. G. Nakagawa, K. Miura, S. Sasaki, and M. Yano, Lens-coupled laser diode module integrated on silicon platform, *J. Lightwave Technol.*, 14, 1519 (1996).

53. S. Sasaki, G. Nakagawa, K. Tanaka, K. Miura, and M. Yano, Marker alignment method for passive laser coupling on silicon waferboard, *IEICE Trans. Commun.*, E79-B, 939 (1996).

54. S. Tsuji, R. Takahashi, T. Kato, F. Uchida, S. Kikuchi, T. Hirataka, M. Shishikura, H. Okano, T. Shiota, and S. Aoki, Passive coupling of a single mode optical waveguide and a laser diode/waveguide photodiode for a WDM transceiver module, *IEICE Trans. Commun.*, E79-B, 943 (1996).

55. M. Gruber, S. Sinzinger, and J. Jahns, Optoelectronic multi-chip-module based on planar-integrated free-space optics, *Proc. SPIE*, 4089, 539 (2000).

56. S.H. Song, C.D. Carey, D.R. Selviah, J.E. Midwinter, and E.H. Lee, Optical perfect-shuffle interconnection using a computer-generated hologram, *Appl. Opt.*, 32, 5022 (1993).

57. A.K. Ghosh, M.B. Lapis, and D. Aossey, Planar integration of joint transform correlators, *Electron. Lett.*, 27, 871 (1991).

58. S. Reinhorn, Y. Amitai, and A.A. Friesem, Compact planar optical correlator, *Opt. Lett.*, 22, 925 (1997).

59. W. Eckert, V. Arrizon, S. Sinzinger, and J. Jahns, Compact discrete correlators with improved design, *Opt. Commun.*, 186, 83 (2000).

60. S.H. Song and E.-H. Lee, Parallel detection of WDM packet addresses by using three-dimensional planar integrated optics, *IEEE Photon. Technol. Lett.*, 9, 112 (1997).

61. S.H. Song, S. Park, E.-H. Lee, P.S. Kim, and C.H. Oh, Planar optical implementation of multichannel fractional Fourier transforms, *Opt. Commun.*, 137, 219 (1997).

62. M. Gruber, J. Jahns, and S. Sinzinger, Planar-integrated optical vector-matrix multiplier, *Appl. Opt.*, 39, 5367 (2000).

63. S.H. Song, J.-S. Jeong, and E.-H. Lee, Beam-array combination with planar integrated optics for three-dimensional multistage interconnection networks, *Appl. Opt.*, 36, 5728 (1997).

64. F.B. McCormick, T.J. Cloonan, A.L. Lentine, J.M. Sasian, R.L. Morrison, M.G. Beckman, S.L. Walker, M.J. Wojcik, S.J. Hinterlong, R.J. Crisci, R.A. Novotny, and H.S. Hinton, Five-stage free-space optical switching network with field-effect transistor self-electro-optic-effect-device smart-pixel arrays, *Appl. Opt.*, 33, 1601 (1994).

65. L.A. D'Asaro, L.M.F. Chirovsky, E.J. Laskowski, S.S. Pei, T.K. Woodward, A.L. Lentine, R.E. Leibenguth, M.W. Focht, J.M. Freund, G. Guth, and L.E. Smith, Batch fabrication and testing of GaAs-AlGaAs field effect transistor self electro-optic effect device (FET-SEED) smart pixel arrays, *IEEE J. Quant. Electron.*, 29, 670 (1993).

6

Optical Bus Technology

HIDENORI YAMADA, TAKESHI NAKAMURA,
OSAMU UENO, and TAKESHI KAMIMURA

CONTENTS

6.1 INTRODUCTION

In ordinary homes, practical use of the technology relevant to computers and networks, including the Internet, is increasing quickly with the rapid spread of high-speed and mass networks, such as ADSL (asymmetric digital subscriber line), CATV (cable antenna television), and FTTH (fiber to the home). The demand for the processing speed of a system is increasing with the spread of the Internet and multimedia. Although the central processing unit (CPU) acceleration every year has corresponded to this increase, the data transmission speed of the electric wiring in the equipment is a bottleneck, and the processing speed of a system is restricted. Some serious problems are occurring in electric wiring.

1. Fan out of high speed signals are restricted by wiring length. Moreover, wiring area and power consumption are increasing.
2. As driving frequency gets high, wiring on printed wiring board becomes difficult, so as to equalize wiring length.
3. As the number of LSI high speed I/O pin increases, wiring on printed wiring board becomes complicated. This results in increase of EMI (electromagnetic interference) noises.

When transmitting information, optical interconnection technology solves these problems, and is attractive technology. However, in order to use optical interconnection technology inside equipment, their costs and mounting cost are still expensive, and are not applicable to the equipment for industry or for office use not to mention home use. And it was difficult to realize the interconnection of many nodes with simple composition.

To solve these problems, we developed the technology called an optical sheet bus (OSB) using diffusion transmissions of light. This chapter explains the principle and function of OSB in order of the package of the element of optical part, a control circuit, and an application system following the beginning.

6.2 OPTICS IN OPTICAL BUS TECHNOLOGY

6.2.1 Introduction

Optical interconnections are one of the most promising technologies to relieve the bottleneck of electronic interconnections. After the historical paper on optical interconnections by Goodman,[1] many optical interconnection technologies were developed and reported.[2] Examples include a planar waveguide with refractive and diffractive optics,[3] a bidirectional linear optical backplane,[4] and a two-dimensional (2D) planar waveguide.[5]

However, most of these technologies are for point-to-point communications, and it is difficult to realize communications between multiple nodes with high distribution uniformity. Communications between multiple nodes are popular in electronic buses, and an optical bus with such ability has been expected to replace electronic buses.

6.2.2 Optical Sheet Bus with Diffused Light Transmission

6.2.2.1 What the OSB Offers

We proposed novel OSB interconnection technology that enables realization of communications between multiple nodes with high distribution uniformity utilizing diffused light transmission in planar waveguides.[6] The OSB does not need high accuracy like conventional waveguides or optical fibers, so the OSB offers low-cost optical buses and links with the advantages mentioned here.

6.2.2.2 Basics of the OSB

An OSB consists of one or more planar lightguides and diffuser (Figure 6.1). A planar lightguide is made of a core layer and claddings. The core and claddings can be made of transparent

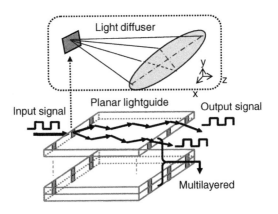

Figure 6.1   Principle of the OSB.

material such as optical plastics or glass, and the claddings have a lower refraction index than the core. A typical example of core material is PMMA (polymethylmethacrylate), with a refractive index of 1.49; claddings of perfluorinated polymer have a refractive index of 1.34. The lightguide has optical surfaces except for the input surface. The input surface has a diffusing structure or one or more diffuser elements on it.

The input optical signal is diffused and transmitted through the planar lightguide, and it exits from the output surface. The diffuser has low insertion loss, high uniformity, and an optimized diffusing angle for an OSB lightguide.

Major advantages of the OSB are the following:

1. Broadcasting.
2. Large fan out derived from high distribution uniformity in output nodes.
3. Wide alignment tolerance.
4. Bidirectional communication between multiple nodes.
5. Ability of combiner suitable for multiplexed communications.
6. Multibit transmission with layered structures.

Figure 6.2 shows a basic image of an OSB applied to a CPU–memory bus. Both the CPU module and the memory modules have light sources and photoreceptors and communicate

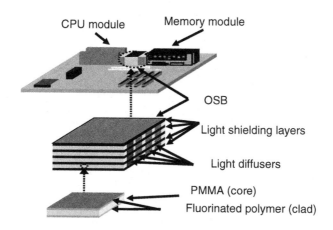

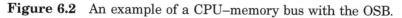

Figure 6.2 An example of a CPU–memory bus with the OSB.

with each other with the OSB lightguide. There are light-shielding layers between four OSB lightguides.

6.2.2.3 Broadcasting and Fan Out

Because signal lights in the OSB are diffused and uniformly distributed at the output surface, the OSB can broadcast signals. Dimensions of the OSB lightguides are designed to make the light at the exit surface uniform with internal reflection.

We verified the broadcasting of the OSB with a lightguide. The refraction index of the PMMA core layer was 1.49; the cladding layer was the air, which had an index of 1.0. The lightguide dimensions were 4.2 mm wide (x-axis in Figure 6.1), 15.2 mm long (z-axis in Figure 6.1), and 1.0 mm thick (y-axis in Figure 6.1). Diffusing angles from the diffuser are 60° full width at half maximum (FWHM) in the x–z plane in Figure 6.1 and 10° FWHM in the x–y plane. Measured uniformity of the light from the exit surface of the OSB lightguide was as low as 3%. Uniformity is defined by the equation (Max – Min)/(Max + Min) × 100(%). Maximum fan out at 500 Mb/sec can be 30 or more as described in section 6.2.3. This is difficult with electronic interconnections at this high bit rate.

6.2.2.4 Wide Alignment Tolerance

Because output signals from OSB lightguides have high uniformity in optical power, position errors of photodetectors at the output nodes of the OSB and the lightguide itself hardly influence output signal levels from the photodetectors. Thus, the OSB can widen alignment tolerances of various parts used in the OSB.

Figure 6.3 shows the ability of the OSB widening alignment accuracy of various parts based on simulations. The plotted line at an incident angle of 0° perpendicular to the incident surface and that at an incident angle of 3° are almost the same as shown in Figure 6.3a. The result with the light source positioned at the optical axis and that with the light source shifted from the optical axis by 0.3 mm are also almost the same as shown in Figure 6.3b. In all the cases, the calculated intensity of light was uniform within –0.3 to 0.3 mm in the

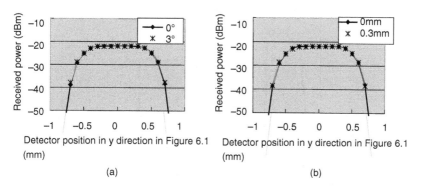

Figure 6.3 Alignment tolerance to (a) incident angle and (b) incident position calculated by ray tracing.

y-axis in Figure 6.1. This means that alignment tolerances of the light source, lightguide, and photodetector can be up to 3° or 0.3 mm in this case. One cannot achieve these tolerances with conventional lightguides or fibers.

The dimensions of the OSB lightguide used here were as follows: 40 mm width, 40 mm length, and 1 mm thick. Diffusing angles from the diffuser were 90° FWHM in the x–z plane in Figure 6.1 and 30° FWHM in the x–y plane. The diameter of the photodetector used in the simulations was 0.4 mm.

6.2.2.5 Bidirectional Communication between Multiple Nodes

We made various types of optical buses that enable bidirectional communications between multiple nodes as described in sections 6.2.4 and 6.5.1. One example is the one-layer OSB bus with a transmission rate of 100 Mb/sec. Cross talk between the two directions was around 16 dB.

6.2.2.6 Multibit Transmission

An example of multibit transmission buses we have made is a three-layered OSB bus. A clock signal, a reset signal, and a serialized signal from four-bit parallel signals were transmitted through the three-layer OSB. The bit rate of the original

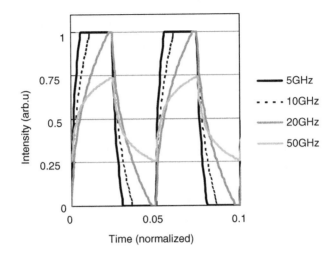

Figure 6.4   Transmission bandwidth of an example of the OSB.

parallel signal was 192 Mb/sec, and that of the serialized one was 768 Mb/sec.

6.2.2.7 Bandwidth

Figure 6.4 shows the simulated result of transmitted clock signals with a 40-mm long OSB lightguide. The PMMA core of the lightguide had a refractive index of 1.49, and the perfluorinated polymer cladding had a refractive index of 1.34. The bandwidth of the OSB lightguide shown here was around 20 Gb/sec.

Figure 6.5 shows an eye diagram at 2 Gb/sec obtained by an experiment with a 2-mm long OSB lightguide. The eye opened widely enough to transmit signals in good condition. The bit rate in this experiment was limited by the photodiode (PD) used.

6.2.3   Star Coupler

The OSB enabled a novel star coupler with high uniformity and low excess loss at a low cost.[7] An OSB star coupler consists of an OSB lightguide and plastic optical fibers (POFs) as shown in Figure 6.6.

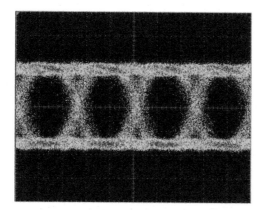

Figure 6.5 An eye diagram at 2 Gb/sec.

Measured properties of three types of OSB star couplers (4 × 4, 8 × 8, and 16 × 16) are shown in Figure 6.7. The core of the fabricated lightguide was made of PMMA, and the air was the cladding. The maximum distribution ratio was as low as 0.8 dB; excess losses are almost constant with the increase of the number of branching, as shown in Figure 6.7b. This excellent property was achieved by diffused light transmission and is difficult to accomplish other than by OSB technology.

The distribution ratio and excess loss are defined as follows:

Distribution ratio (dB) = Maximum insertion loss − Minimum insertion loss

$$\text{Excess loss (dB)} = -10 \log(\textstyle\sum P_{out}/P_{in})$$

P_{in} is the input optical power, and P_{out} is the output optical power.

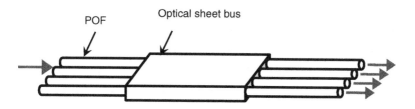

Figure 6.6 A star coupler using the OSB.

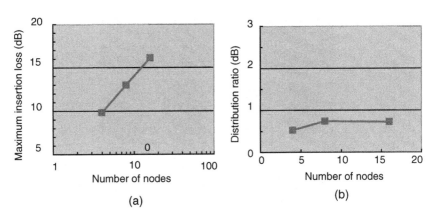

Number of nodes

(a)

Number of nodes

(b)

Figure 6.7 (a) Maximum insertion loss and (b) distribution ratio of OSB star couplers.

Figure 6.8 shows the relationship between bit error rate (BER) and received optical power at 100, 400, and 500 Mb/sec. The optical power needed at a BER of 10^{-11} at 500 Mb/sec is −20.5 dBm. The minimum output power from a 16 × 16 OSB star coupler is −15.8 dB with a 0-dBm light source as shown

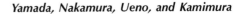

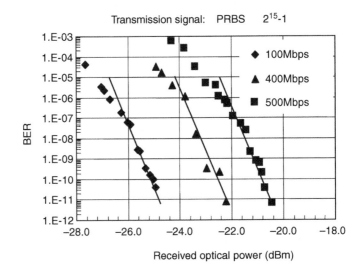

Figure 6.8 BER vs. received optical power.

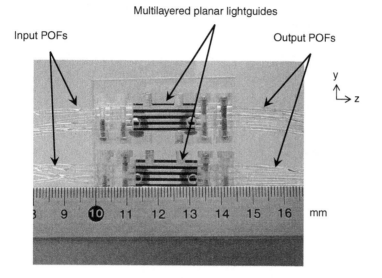

Figure 6.9   A photograph of a multilayer OSB star coupler.

in Figure 6.7. This is much larger than the needed power at the BER of 10^{-11} at 500 Mb/sec. Because the margin is 4.7 dB, the maximum branching number would be around 32.

OSB couplers can be easily stacked into a multilayer structure as shown in Figure 6.9. A stacked OSB coupler requires less space than combining multiple couplers and is suitable for multibit buses.

The OSB star coupler is remarkably cost-effective because it can be mass-produced by injection molding.

6.2.4   Optical Backplane

We have developed two types of optical backplanes using the OSB. One uses OSB star couplers for signal distribution, and the other uses a new type of step-shaped planar lightguides.

Figure 6.10 shows an example of an optical backplane bus using the OSB star couplers. An eye diagram obtained by this example opened wide enough to transmit signals in good condition.

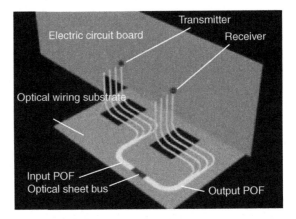

Figure 6.10   An optical backplane bus using OSB star couplers.

Figure 6.11 shows an example of the new step-shaped planar lightguides. The size of the lightguide is 4 × 45 × 1 mm. The newly developed backplane has a slanted surface at each input and output node. Because the new backplane lightguides (1) need no fiber wiring, (2) reduce the number of parts, (3) enable easy assembly, and (4) can be mass-produced by injection molding, they can greatly reduce the cost of optical backplanes. An array of this type of lightguide realizes a multibit bus. Experimental results with the lightguide in Figure 6.11 were as follows: Maximum insertion loss was as low as 8.5 dB, and the distribution ratio was as low as 0.4 dB.

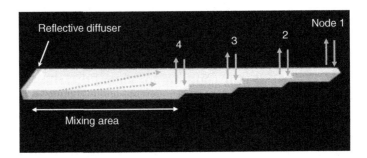

Figure 6.11   Schematic of the new step-shaped planar lightguides.

We will show an example application to take advantage of this new lightguide in section 6.5.1.1.

6.3 OPTICAL DEVICES AND THEIR PACKAGING DESIGN

6.3.1 Vertical Cavity Surface-Emitting Lasers for OSB

Vertical cavity surface-emitting lasers (VCSELs) have been widely recognized as light sources for optical data communications, such as Gigabit Ethernet (GbE) and Fiber Channel. The features of VCSELs are 2D integrability and their low-cost nature, and they have become the *de facto* standard for data communication modules. A more general and detailed description of VCSELs can be found in chapter 1.

In this section, we describe the design and characteristics of Fuji Xerox's VCSELs and their specific features applied to OSB modules.

6.3.1.1 Characteristics of Fuji Xerox VCSELs

Fuji Xerox has commercialized VCSELs since 2001. The features of Fuji Xerox's VCSELs were developed and obtained mainly through development of the light source for laser printers, which require severe and tight specifications for VCSELs, such as high single-mode power and uniformity. The technology developed for the printer light source has been applied for VCSELs for other applications, including data communications and OSBs.

The characteristics of Fuji Xerox's VCSELs are as follows:

- Low operating current: Using an oxide-confined structure, the threshold current of Fuji Xerox's VCSELs is as low as 1 mA for multimode operation. Typical operating current is less than 5 mA for 2 mW of output power.
- High uniformity: High uniformity of device characteristics is achieved utilizing an original oxidation process control called OPTALO, which precisely detects

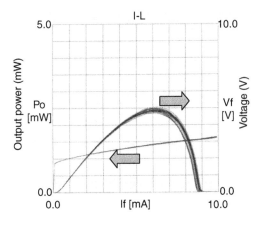

Figure 6.12 Drive current–light output power (I–L) and drive current–drive voltage (I–V) characteristics of 6∗6 VCSEL.

the end point of AlAs oxidation *in situ*. Several other schemes are also incorporated in the device designs and processes (Figure 6.12).

- Large-scale integration: Because of the high uniform characteristics and processes, very large-scale integration of VCSELs was demonstrated using 12 × 120 2D arrays (Figure 6.13).
- High-frequency response: The 10-Gb/sec operation has been achieved with high reliability and low spectral line width required for 10-GbE specifications. The phase control method using a metal aperture was adopted to realize 10-Gb/sec operation (Figure 6.14).
- Low divergence angle: Stable and low divergence angles have been achieved using the same metal aper-

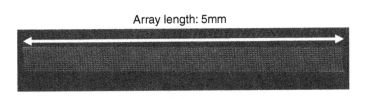

Figure 6.13 A 12 × 120 matrix VCSEL array. White spots show each VCSEL.

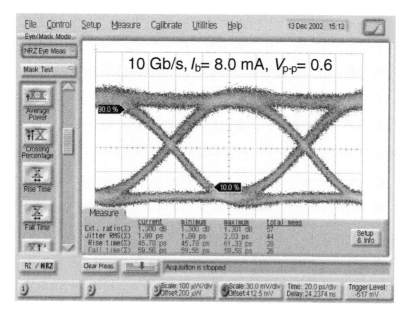

File Control Setup Measure Calibrate Utilities Help   13 Dec 2002 15:12

10 Gb/s, I_b= 8.0 mA, V_{p-p}= 0.6

Figure 6.14   Eye diagram of 10-Gb/sec operation.

ture structure mentioned above, which leads to higher coupling efficiency to optical fiber without any special optics (Figure 6.15).

- High power: Output power as high as 30 mW from a single chip is obtained by densely integrating several VCSELs on one chip, which leads to the adoption of VCSELs to new applications, such as free-space optical communications.

6.3.1.2 VCSEL Characteristics Required for OSB Application

In the bus system that handles a large number of data lines inside individual equipment, the number of VCSELs used in the system becomes large; hence, the VCSELs should satisfy the following characteristics:

- Low driving current capable of complementary metal-oxide semiconductor (CMOS) direct drive to eliminate

– Injection current : 7 mA (constant)   Output power : ~2 mW –

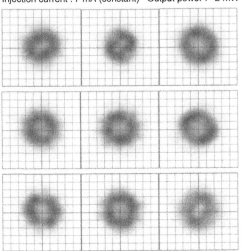

Figure 6.15   Variation of far-field pattern within 3-in wafer (minimum 18°, maximum 22°).

the need for special driver application-specific integrated circuits ASICs.

• Small variations in device characteristics to simplify or eliminate the signal correction circuit.

• Higher power compatible with the high absorption rate media used in OSBs.

In particular, because the OSB uses plastic sheets that have strong absorption in the infrared region, VCSELs for OSB must have as low wavelength and high power as possible to compensate strong absorption within the plastic sheets. We have developed multispot 780-nm VCSELs for OSBs that can emit a few milliwatts output power without sacrificing the lifetime of the devices (Figure 6.16).

6.3.2   PDs for OSBs

A PD for OSBs is selected from the values of the spectrum response range, the cutoff frequency, the dark current, the reverse voltage, and so on according to the wavelength of the light source, the maximum data rate, the minimum input

Figure 6.16 Multispot VCSEL.

power, and other product requirements. Light at the visible range to the near-infrared range is used in many examples of OSB application. Si PIN PDs or GaAs PIN PDs are usually used for these applications. To tolerate displacement of optical components, the active area of the PD has to be large enough. Because the capacitance of Si PDs is smaller than that of the GaAs PD, it is easy to obtain a relatively large active area using Si. GaAs has higher mobility than Si. The Si PD is usually used for OSB applications at less than 1 Gb/sec; the GaAs PD is used for applications that require higher speed.

6.3.3 Packaging Design around Optical Devices

Many OSB applications require that laser diodes (LDs) and PDs are placed close to each other. This section describes the design concept if packaging density is relatively high (Figure 6.17).

The LD and laser diode driver (LDD) are emission sources of electromagnetic noise. The PD and preamplifier (PA) are parts with low immunity to electromagnetic (EM) noise, shown as EM victim in Figure 6.17. When the LD and PD are adjacent, the electromagnetic isolation of these optical parts is important. The first step for the isolation is to decrease the noise energy of the emission source. Selecting an LD with low drive current, such as a VCSEL, and selecting an LDD with low

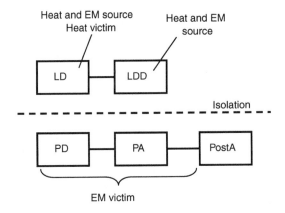

Figure 6.17 Electromagnetic cross talk and thermal cross talk around optical devices. Post A, post-preamplifier.

drive voltage are effective for decreasing the source energy. The next step for the isolation is to cut off the coupling pass. Power planes of the LD and PD in the circuit board are to be separated in many cases. Sometimes, the ground planes of the LD and PD are separated or are shielded individually. The other method for isolation is to minimize antenna elements. The length of the trace between the LD and the LDD, which acts as a transmission antenna, and the length of the trace between the PD and the PA, which acts as a receiving antenna, should be minimized.

Thermal design is also important in packaging the OSB. The LD and LDD are heat sources, and the LD is susceptible to heat. It is effective for the thermal design to decrease the source power and to isolate the source and the victim, which are similar methods to decrease electromagnetic coupling (Figure. 6.17).

6.4 OPTICAL INTERCONNECTION CONTROL CIRCUIT

For optical interconnections, customers expect communications at high speed and long distance with no transmission errors. The signal quality in optical interconnections must be equivalent to the quality in electrical interconnections.

In section 6.4.1, we describe how to achieve the optical bus/ coupler signal quality specifications.

The OSB has the following unique features not included in electrical interconnections:

- No wiring is required to interconnect nodes. Transmission channels for each direction can be flexibly assigned as necessary.
- Multiplexing methods, such as wavelength multiplexing, polarized light multiplexing, and amplitude multiplexing, are possible.
- Transmission direction multiplexing is also possible.

It is important for effective data communications to decrease unused transmission channels while operating. In section 6.4.2, we describe the OSB bus/coupler arbiter that makes good use of the features above and greatly improves transmission efficiency.

6.4.1 Optical Bus/Coupler Signal Quality

In optical signal transmission using the OSB, multiple optical signals flow into each other in a planar lightguide. Thus, the following technologies are indispensable to achieve high-quality OSB bus/coupler signals: direct current (DC) balancing, error correction, and timing adjustment.

6.4.1.1 DC Balancing

A DC balancing method such as 8B10B[1] is already used in conventional optical fiber communications. It becomes more important in multiple point-to-point optical interconnections because each receiver must receive wide-range signals generated by the different transmitters, transmitting paths, optical component implementations, and so on. The signals can be compensated by DC balancing methods.

A problem in 8B10B is to increase transmission latency by processing time for both encoding and decoding. In some applications, such as data transmission between CPU and cache memory, for which low latency is required, a quicker DC balancing method is necessary.

6.4.1.2 Error Correction

The BER specification standard is 1e-12 or better in Fiber Channel standardization FC-1 layer. But BER 1e-12 is not enough for optical interconnection applications. For instance, transmission error occurs every 16 min on 1-Gb/sec transmission lines with a BER that is 1e-12.

If a forward error correction method that automatically corrects 1 bit error in 10 bits/word is applied, the BER is greatly improved to 1e-19. A transmission error occurs every 317 years, and thus error-free transmission is possible. Also, the BER while the system is operating is approximately measured by counting corrected errors. Signal quality can be kept high by using both error correction and the timing adjustment method described next.

Latency, increased by error correction as well as DC balancing, is also a problem in applications for which low latency is required. The latency specification is one of the important points to consider in choosing an error correction method.

6.4.1.3 Timing Adjustment

In high-speed signal transmission either by electrical interconnections or optical interconnections, the BER is affected by transmission signal timing transitions. These are caused by characteristic transitions of interconnection components, environmental variations, device degradation, and so on. In addition, for bus/couplers, transmitter variations must be considered.

The clock data recovery (CDR)[3] method is generally used in point-to-point communications to adjust transmission timing. In a CDR-based receiver circuit, the clock signal is extracted from the data signal, and both signals are synchronized.

If CDR is applied to bus signal transmission as shown in Figure 6.18, the data transmission breaks, and the data signal and clock signal are synchronized whenever the transmitter node is changed. Therefore, CDR can only be used in application systems that allow the transmission break period. Other synchronization methods between clock signal and data

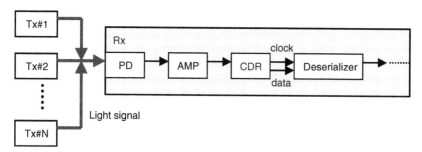

Figure 6.18 Serialized bus system example using CDR.

signal and deskewing methods between multiple data signals are also important in optical bus/couplers.

Each technological method should be determined according to application specifications, such as transmission speed and latency. We actually implemented these methods on application systems mentioned in section 6.5.

6.4.2 OSB Bus/Coupler Arbiter

Figure 6.19a shows an example of signal flow in a memory read cycle, and Figure 6.19b shows signal flow in a memory write cycle. You can see that the read/write operation requires different bandwidth in each transmitting direction. If a transmission channel for each direction can be flexibly used as necessary, memory access is faster because unused transmission channels are reduced.

We define transmission efficiency E as follows:

$$E = 1 - U/T$$

Here, T is the total bandwidth of the interconnection, and U is the total bandwidth of unused transmission channels.

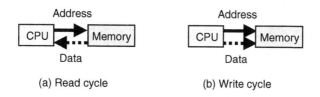

Figure 6.19 Signal flow in memory access cycle.

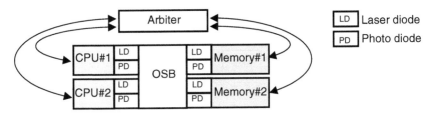

Figure 6.20 A multiprocessing system example including an OSB and an arbiter.

The purpose of the OSB arbiter is to improve the efficiency E in OSB-based systems. Figure 6.20 shows a system example that includes an OSB that operates as a 2×2 coupler and an OSB arbiter. It is a multiprocessing system in which the OSB interconnects two CPU nodes and two memory nodes with multiple transmission channels. The system enables six patterns of memory access (Figure 6.21).

Figure 6.22 shows the block diagram of the OSB arbiter. It is connected with all nodes to accept each transmission request and to inform each arbitration result. Each node is

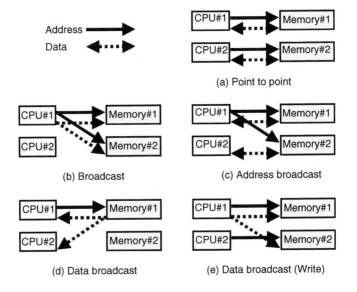

Figure 6.21 Access patterns between two CPUs and two memories.

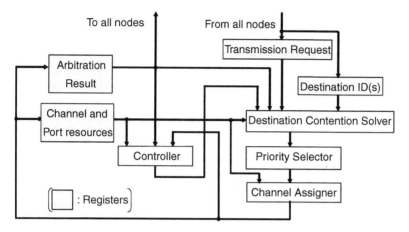

Figure 6.22 Arbiter block diagram.

discriminated by an ID number. The following are arbiter input signals asserted by CPU nodes:

- Transmission request.
- Destination node ID(s) for forward broadcasting.
- Destination node ID(s) for backward broadcasting.

Here, forward broadcasting is broadcasting from CPUs to memories, such as in Figures 6.21b, 6.21c, and 6.21e. Backward broadcasting is broadcasting from memories to CPUs, such as in Figure 6.21d. Considering the access priority of each node, the arbiter first solves destination node contentions for each direction and then solves transmission channel contention for each direction. Finally, the OSB arbiter assigns transmission channels to the "winner" nodes and updates access priorities of all nodes for fair assignment. The following are the arbiter output signals:

- OK/NG (whether each node can transmit).
- If OK, transmission channel number(s) allocated for forward and backward broadcasting.

It is important that the assigned transmission channels are never unused, thus improving transmission efficiency. We developed the OSB arbiter for 3 × 3 optical couplers by a complex programmable logic device (CPLD) chip using

6000 gates. In some application systems, the 65% channel efficiency of the conventional arbiter was improved to 90% by the OSB arbiter.

6.5 APPLICATION SYSTEMS

Optical interconnection technology mainly was introduced in interchassis interconnections, but it also has been desired intra-chassis as high-speed and EMI-free board-to-board in interconnections. The following two problems were barriers to the introduction of optical interconnections inside the chassis:

- Optical components are expensive to apply inside the chassis.
- Optical signal quality must be equivalent to electrical signal quality, with no manual adjustment for interconnections.

We have mentioned solutions for the second problem in section 6.4.1.

Interconnections inside the chassis often require multiple board-to-board interconnections. As for the first problem, optical components are expensive if only point-to-point interconnections by optical fibers are used to build multiple board-to-board interconnections. For instance, as shown in Figure 6.23b,

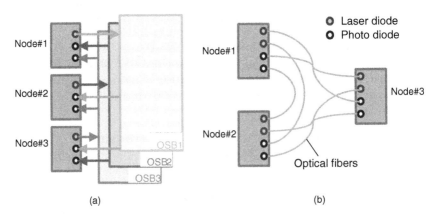

(a)   (b)

Figure 6.23 Comparison of OSB and optical fibers on nonblocking switching systems for three nodes.

a nonblocking switching system with N input/output (I/O) nodes based on optical fibers needs $2_N C_2$ ($=N(N-1)$) optical fibers, $N(N-1)$ LDs, and $N(N-1)$ PDs. On the other hand, the same system using an OSB includes only N lightguides, N LDs, and $N^2 - N$ PDs (Figure 6.23a). It can be seen that low-cost optical bus/couplers can solve the first problem by reducing the number of optical components.

The same problem appears in electrical interconnections. If a transmission speed of more than a few hundred megabits per second is required, only point-to-point transmission is possible in electrical interconnections. This results in a system with many copper wire cables and transceiver circuits.

Now, optical interconnections are introduced in applications such as factory automation systems, central exchange systems for cellular phone/telephone lines, emulators, and video-processing systems. In the following sections, we mention two interesting application system examples of board-to-board optical interconnections.

6.5.1 Optical Backplane Application

In factory automation industries, sensor signal-processing equipment is widely used. It has multiple input/output (I/O) nodes for data acquisition and CPU nodes for data processing. To connect I/O nodes and CPU nodes, scalable interconnections such as backplanes are desired because the number of nodes varies depending on each customer's requirement.

However, backplane signal transmission error caused by EMI from neighboring mechanical components and the like is a big problem, which is why optical backplanes are desired to transmit signals at several hundred megabits per second.

We developed an optical backplane prototype system as shown in Figure 6.24 and Figure 6.25. The system consists of 11 application boards, which are 1 CPU board and 10 I/O boards, 1 optical backplane, and 11 optical interface boards that connect each application board with the optical backplane. The 28-b and 33-Mb/sec electrical signals are serialized and converted into 6-b and 320-Mb/sec optical signals

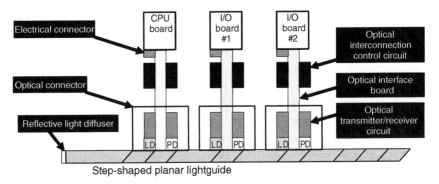

Figure 6.24 Optical backplane-based system prototype.

in the optical interface board. The following are the principal features:

- The optical backplane includes six lightguides and some electrical wiring, such as power supply lines, and some low-speed control signals.
- The lightguide is a step-shaped planar type with 11 nodes. (Figure 6.25 shows the same type of lightguide with 8 nodes.) It can be easily fabricated by injection molding.
- An optical connector is mounted on each optical interface board to connect with the optical backplane. It includes VCSELs, VCSEL drivers, PDs, and transimpedance amplifiers for 6-b optical I/Os.

Figure 6.26 shows that eye diagrams of a transimpedance amplifier output signal of 320 Mb/sec have no problems. The CPU board can access each I/O board successfully.

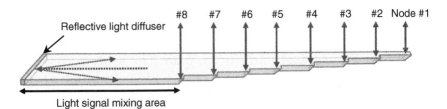

Figure 6.25 Shown is 1 b of a step-shaped planar lightguide with eight I/O nodes.

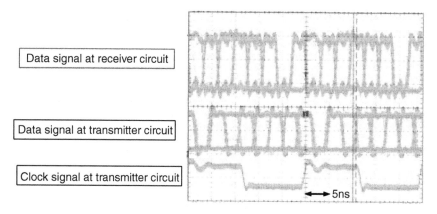

Figure 6.26   The 320-Mb/sec signal waveforms in an optical back-plane prototype system.

We compare the number of optical components required to build an OSB-based prototype system with that of an optical-fiber-based system in Table 6.1. The OSB is effective in optical backplane cost reduction.

6.5.2   Optical Coupler Application

The in-circuit emulator (ICE) system is a well-known tool for microprocessor development. It is especially used to develop embedded microprocessors. It is implemented by electric circuits on printed-wiring boards (PWBs), and precisely works the same as the target microprocessor. Microprocessor developers have two important reasons to develop an ICE system: to verify the logic circuits of the target microprocessor before

TABLE **6.1**   Number of Optical Components Required in an Optical Backplane System

	Number of lightguides or fibers	Number of laser diodes	Number of photodiodes
Optical-sheet-bus-based system	7	67	76
Optical-fiber-based system	120	120	120

TABLE 6.2 Primary Specifications of In-Circuit Emulator (ICE)
Prototype System Using Optical Sheet Bus (OSB) Couplers

Target CPU	Panasonic MN103S
Primary optical signals	Data 8 b, address 10 b, control 4 b, clock 1 b
Transmission speed	100 Mb/sec
Number of Nodes	9 (1 CPU node and 8 memory nodes)
Lightguide	1 × 8 OSB coupler
VCSEL/PD packages	5-b array package each
Optical connector	Including VCSEL/PD packages and amplifiers

Note: CPU, central processing unit; PD, photodiode; VCSEL, vertical cavity surface-emitting laser.

fabrication and because the developers need a software development tool before the target microprocessor is completed.

A CPU–memory bus is used in the ICE system. The problem is that the bus transmission speed decreases as the fan-out number increases. This is because the transmission signal rise/fall time lengthens by incremental electrical capacity and inductance.

We applied optical couplers to an ICE prototype. Here, there was only one CPU, and no data transmission between memories was necessary. Thus, we applied optical couplers, not optical backplanes.

The primary specifications, overview pictures, and eye diagrams are shown in Table 6.2, Figure 6.27, and Figure 6.28, respectively. One of the features of the prototype is direct VCSEL driving by transistor-transistor logic (TTL)/CMOS logic circuits. This results in low latency and a low-cost system with no VCSEL drivers.

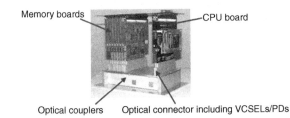

Memory boards   CPU board

Optical couplers   Optical connector including VCSELs/PDs

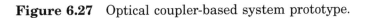

Figure 6.27 Optical coupler-based system prototype.

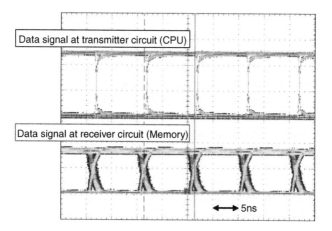

Figure 6.28 The 100-Mb/sec signal waveforms in an optical coupler prototype system.

The ICE prototype with eight memory nodes operates at a 100-Mb/sec transmission speed, which is two times faster than a conventional electrical bus.

REFERENCES

1. J.W. Goodman, Optical Interconnections for VLSI Systems, *Proc. IEEE*, 72, 850 (1984).

2. C. Tocci and J. Caulfield, Eds., *Optical Interconnection— Foundation and Applications*, Artech House, Norwood, MA, 1994, 65.

3. J. Jahns and A. Huang, Planar integration of free-space optical components, *Appl. Opt.*, 28, 1602 (1989).

4. S. Natarajan et al., Bi-directional optical backplane bus for general purpose multi-processor board-to-board optoelectronics interconnects, *J. Lightwave Technol.*, 13, 1031 (1995).

5. S. Kawai and M. Mizoguchi, Two-dimensional optical bus for massively parallel processing, *Jpn. J. Appl. Phys.*, 31, 1663 (1992).

6. M. Funada et al., Proposal of novel optical bus system using optical sheet bus technology, *SPIE Proc.*, 3632, 30 (1999).

7. O. Takanashi, High-uniformity star coupler using diffused light transmission, *IEICE Trans. Electron.*, E84-C, 339 (2001).

7

Fiber Optics

NAOYA UCHIDA

CONTENTS

7.1 INTRODUCTION

More than 20 years have passed since optical fibers were introduced in commercial transmission systems. As shown in Figure 7.1, the fiber has various advantageous features with a strong impact on transmission systems. Thus, the systems receive a number of advantages superior to those of conventional transmission systems. Development of optical transmission systems has been smooth, and it is doubtless that the fibers played an indispensable role in the realization of the information society age.

In this chapter, I describe basic features of the fiber: structure, fabrication, transmission characteristics, mechanical properties, and reliability assurance. Requirements for fibers used in wavelength division multiplexing (WDM) transmission systems are discussed, and novel fibers developed for WDM transmission systems are described in detail.

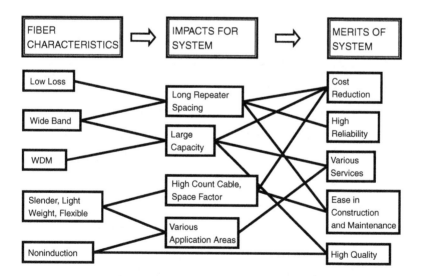

Figure 7.1 Characteristic features of optical fibers, their impacts on optical transmission systems, and the resultant advantages of the systems. WDM, wavelength division multiplexing.

7.2 BASIC PRINCIPLES OF OPTICAL FIBERS

7.2.1 Fiber Structure

As shown in Figure 7.2a, an optical fiber is composed of the circular core and cladding. The refractive index of the core is set larger than that of the cladding, and an optical wave propagated along the fiber is effectively confined in the core region. On the basis of geometric optics, an optical wave is propagated and repeats total internal reflection at the core–cladding boundary (Figure 7.2b).

Figure 7.3 is a schematic illustration of fiber types. The fiber is classified into two categories: multimode fiber and single-mode fiber (SMF). When the core diameter or the refractive index difference between core and cladding are large, a plural number of modes can be propagated in the fiber; such a fiber is called a multimode fiber. When the core diameter or the refractive index difference becomes small, the number of propagation modes decreases, and finally only a fundamental mode can be propagated. Such a fiber is called a single-mode

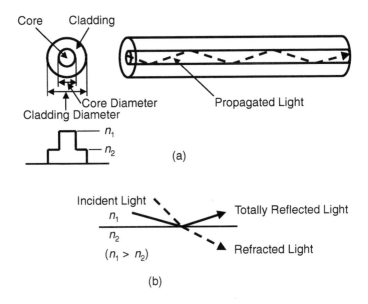

Figure 7.2 Schematic illustrations of (a) optical fiber structure and (b) total internal reflection.

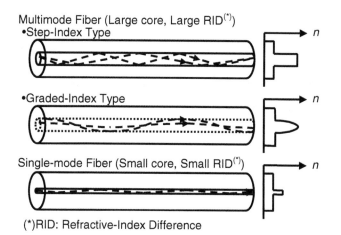

Multimode Fiber (Large core, Large RID$^{(*)}$)
• Step-Index Type

• Graded-Index Type

Single-mode Fiber (Small core, Small RID$^{(*)}$)

(*)RID: Refractive-Index Difference

Figure 7.3   Classification of fiber types in view of refractive index profiles.

fiber. As for the multimode fiber, transmission bandwidth changes markedly by the change in the refractive index profile. The fiber with a graded index (GI) profile has a wider bandwidth than that with a step index profile, and the optimum profile of the GI fiber for attaining the largest bandwidth is nearly parabolic.

7.2.2   Fiber Material

Table 7.1 summarizes typical fiber types classified on the basis of fiber constituent materials. The most widely used fiber is a silica glass fiber, and a few are other types of fibers. At an early stage of fiber research and development, compound glass fiber and plastic-clad fiber were developed in parallel with the

TABLE 7.1   Classification of Optical Fibers in View of Fiber Constituent Materials

Fiber type	Core	Cladding
Silica glass fiber	Silica glass	Silica glass
Compound glass fiber	Compound glass	Compound glass
Plastic-clad fiber	Silica glass, compound glass	Plastic
Plastic fiber	Plastic	Plastic

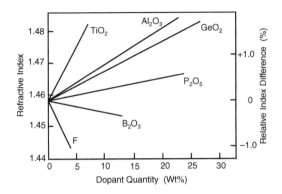

Figure 7.4   Refractive index change of silica-based glass for various dopants.

silica glass fiber. However, because fabrication techniques for the silica fiber were established and inexpensive silica fibers with superior transmission characteristics and high reliability were obtained easily, silica fiber became the unique fiber for use in public communications networks and the majority fiber for use in private networks. Below, I describe only the silica fiber. The plastic fiber is important as a transmission medium in the field of optical interconnection. The recent development of the plastic fibers will be described in chapter 8.

To form a waveguide structure inside a silica fiber, several dopant materials were investigated. Figure 7.4 shows the refractive index change as a function of content quantity for typical materials. The refractive index of silica glass is increased by doping TiO_2, Al_2O_3, GeO_2, and P_2O_5; the index is decreased by doping B_2O_3 and F. Among these materials, GeO_2 is generally used as a dopant for the core region, and F is used sometimes as a dopant in the cladding region.

7.3 HISTORY OF OPTICAL FIBER TECHNOLOGY

Figure 7.5 summarizes important items and events related to development of silica optical fibers from 1966 to the present. In 1966, Kao and Hockham[1] predicted that a glass fiber was a possible candidate for a long-distance transmission medium

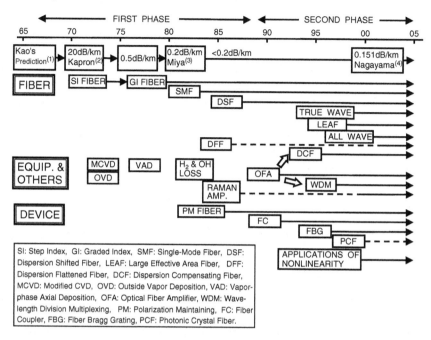

Figure 7.5 Historical flow of development items and events related to silica optical fibers.

for optical waves, with a high potential for low loss and wide bandwidth. A fiber with 20-dB/km loss was achieved by Kapron et al.[2] in 1970. The loss value was epoch-making because it meant that, by using the fiber, a long-distance transmission system longer than 1 km was possible. Extensive research and development began for practical optical fibers. Loss reduction progressed promptly; within ten years, a very low loss of nearly 0.2 dB/km was attained.[3] Recently, an ultimate loss of 0.15-dB/Km was attained.[4]

Figure 7.6 plots a feature of the optical loss reduction by year. Such rapid progress is owed to the exhaustive removal of impurities in fiber (i.e., metal ions and OH ion). This turned the so-called long-wavelength transmission systems utilizing the wavelength regions from 1.3 to 1.6 μm, instead of the previous short-wavelength system utilizing the 0.85 μm wavelength region: The long-wavelength system is decisively advantageous

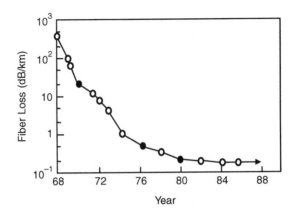

Figure 7.6   Loss reduction of silica fiber year by year.

from the viewpoints of low-loss and low-dispersion characteristics of fiber.

As for multimode fibers, returning to Figure 7.5, a step index fiber was developed first, but soon the step index fiber was exceeded by a GI fiber because of improved controlling techniques for the refractive index profile. At the beginning of the 1980s, commercial optical fiber transmission systems using the GI fiber were realized.

Development of an SMF was expected for a long time because of the ultimate potential for large-capacity transmission. Extensive efforts overcame fabrication and connection technique difficulties, resulting in the commercial use of the SMF in the 1.3-μm wavelength region early in the 1980s.

Utilization of the lowest loss in the 1.55-μm region was strongly desired for high-bit-rate transmission systems, and various attempts were made to shift the zero-dispersion wavelength to the 1.55-μm region. As a result, a convex-index-type dispersion-shifted fiber (DSF) was developed[5] and introduced extensively to long-haul backbone transmission lines in Japan. The period up to the commercial introduction of the DSF is regarded as the first phase of fiber development history.

At the end of the 1980s, an optical fiber amplifier was raised on a practical stage because of the realization of a high-power laser diode (LD) and a highly efficient Er-doped fiber.[6]

The event was also epoch-making, and transmission functions rose to a higher level. Thus, WDM transmission systems became cost-effective for the first time and were introduced mainly in the United States in accordance with the explosive increase in data traffic via the Internet. As a result, novel fibers such as TrueWave® (OFS-Fitel, Inc., U.S.) and LEAF (Large Effective Area Fiber)® (Corning, Inc., U.S.), taking into consideration the fiber dispersion and nonlinear effects, were developed. The period starting from the realization of the optical fiber amplifier is called the second phase of fiber history.

Several fiber-type devices (e.g., polarization-maintaining fiber, fiber coupler, and fiber Bragg grating) were developed, and these devices have been widely used in transmission equipment. Applications of fiber nonlinearity and a novel photonic crystal fiber (PCF) are attractive, and extensive research and development is now in progress.

7.4 MANUFACTURING OF SILICA FIBER

7.4.1 Preform Fabrication

The most important conditions for fiber fabrication are (1) impurity-free fiber, (2) excellent controllability of waveguide structure, and (3) realization of highly reliable fiber with sufficient strength. Various attempts were made to fabricate fiber preform. The following three methods are mainly utilized: modified chemical vapor deposition (MCVD), outside vapor deposition (OVD), and vapor-phase axial deposition (VAD).

7.4.1.1 MCVD Method

Figure 7.7 shows a schematic illustration of the MCVD method.[7] A basic principle of the method is based on chemical vapor deposition (CVD), and a characteristic feature of the method is that fine glass particles formed by the CVD process are sintered immediately to transparent glass films. First, $SiCl_4$ gas for raw material and then $SiCl_4$ gas and a gas for dopant material such as $GeCl_4$ are set to flow with a carrier gas of O_2 into a silica tube. The tube is rotated and is heated by an oxy-hydrogen flame. Oxidization occurs inside the tube, and

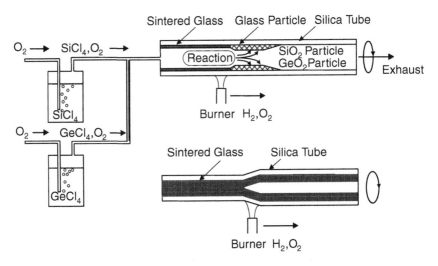

Figure 7.7   Schematic illustration of MCVD process.

SiO_2 or SiO_2–GeO_2 particles are deposited on the inner wall of the tube. The particles are vitrified immediately because of the high-temperature oxy-hydrogen flame (e.g., 1500°C or more), forming transparent SiO_2 or SiO_2–GeO_2 films. A number of film layers are stacked by traversing the oxy-hydrogen flame repeatedly along the rotating tube.

After a desired quantity of glass layers is deposited, the tube is collapsed under the higher temperature condition with the same oxy-hydrogen flame, and a fiber preform is formed. The deposited layers of SiO_2–GeO_2 form a core, those of SiO_2 form an inner cladding, and the collapsed silica tube forms an outer cladding. The refractive index profile of the preform exhibits entirely similar figures to those of the fiber. The most important advantages of the MCVD method are that precise control of the refractive index profile is possible and a complex profile can easily be formed.

7.4.1.2   OVD Method

Figure 7.8 shows a schematic illustration of the OVD method.[8] $SiCl_4$ gas for raw material and that for dopant material, such as $GeCl_4$, are blown on a rotating and traversing ceramic rod

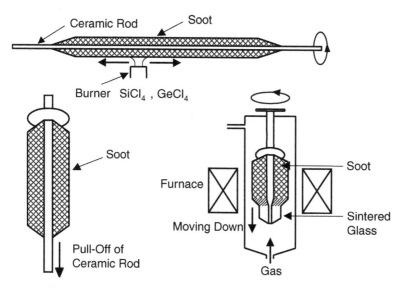

Figure 7.8 Schematic illustration of OVD process.

with an oxy-hydrogen flame, and porous SiO_2–GeO_2 fine particles, called soot, are deposited around the rod. After forming a core part, a cladding part is deposited using $SiCl_4$ gas in a series of continuous processes. When the deposition is completed, the ceramic rod is pulled off, and the remaining soot rod is dehydrated and then sintered into a transparent glass preform in an electric furnace. Advantages of the OVD method are that the soot deposition speed is fast, and essentially a large-size preform can easily be fabricated.

7.4.1.3 VAD Method

Figure 7.9 shows a schematic illustration of the VAD apparatus.[9] The VAD method combines the deposition technique of the OVD method and an axial growth technique suitable for mass production. $SiCl_4$ gas for raw material and that for dopant material, such as $GeCl_4$, are blown in a glass vessel at the lower end of a rotating silica rod with an oxy-hydrogen flame, and SiO_2–GeO_2 core soot is formed at the end of the silica rod. The silica rod is raised gradually, and the core soot is grown in the axial direction. At the same time, $SiCl_4$, O_2, and H_2 gases

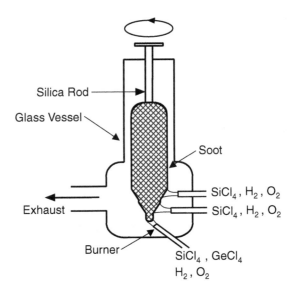

Figure 7.9   Schematic illustration of VAD apparatus.

are blown on the side of the SiO_2–GeO_2 soot from either one or more burners, and soot for an inner cladding region is formed simultaneously. The entire soot rod is pulled up during the deposition process; in principle, a long soot rod can be produced. The dehydrating and sintering processes of the VAD method are the same as those of the OVD method.

The sectional area of the cladding region occupies a major part of the whole cross section for SMF. Therefore, it is disadvantageous to form the whole soot rod inside a glass vessel. Thus, the soot rod consisting of core and inner cladding is once dehydrated and sintered to a transparent rod, and high-speed deposition of an outer cladding is made on the sintered rod in the same manner as in the OVD method.

7.4.2   Fiber Drawing

Figure 7.10 shows a schematic illustration of fiber drawing. A preform rod is inserted in an electric furnace heated above 2000°C, and the lower end of the preform is softened. A fiber with an outer diameter of 125 µm is directly drawn from the preform with a diameter from several centimeters to larger

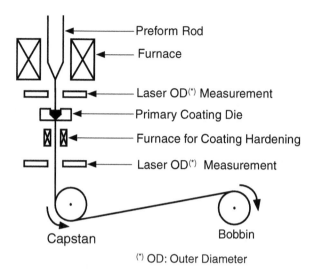

Figure 7.10 Schematic illustration of fiber-drawing apparatus.

than 10 cm. The furnace temperature is precisely controlled because temperature fluctuation causes fiber outer diameter variation. The rotation speed of a capstan is controlled by a monitored signal of the fiber outer diameter, and the fiber outer diameter is automatically held constant. To avoid weakening of fiber strength because of the growth of flaws, the bare fiber is coated with ultraviolet (UV)-curable soft plastic resin immediately below the furnace before the fiber touches the capstan. Also, the atmosphere in the furnace and that surrounding the bare fiber are maintained dust free. Owing to the development of the drawing techniques, a drawing speed faster than 1 km/min is accomplished successfully.

7.5 TRANSMISSION CHARACTERISTICS OF FIBERS

7.5.1 Transmission Bandwidth

7.5.1.1 Multimode Fiber

Geometrical optic treatment gives a simple and plain approximation for optical propagation in fiber. In contrast, wave optic treatment gives more accurate solutions, but analytical

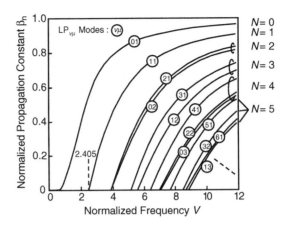

Figure 7.11 Normalized propagation constant β_n as a function of normalized frequency V for the step index fiber. A pair of figures attached to an individual curve corresponds to suffixes $\nu\mu$ of the LP$_{\nu\mu}$ mode.

solutions cannot be obtained for a general shape of refractive index profile in the fiber cross section. Therefore, weakly guiding approximation[10] is used generally to analyze propagation modes when the refractive index difference between core and cladding is small. Figure 7.11 shows dispersion curves of propagation modes for a step index fiber. Here, the horizontal axis represents the normalized frequency, usually called the V number

$$V = (2\pi a/\lambda)\left(n_1^2 - n_2^2\right)^{1/2} \tag{7.1}$$

where a is the core radius, λ is the optical wavelength, and n_1 and n_2 are the refractive indices of the core and the cladding, respectively. The vertical axis represents the normalized propagation constant β_n:

$$\beta_n = \left[(\beta/k)^2 - n_2^2\right]/\left(n_1^2 - n_2^2\right) \tag{7.2}$$

where β is the propagation constant, and k is the wave number, $k = 2\pi/\lambda$. Here, the mode is represented by a linearly polarized (LP) mode LP$_{\nu\mu}$, which is a superposition of strict vector electromagnetic modes, HE, EH, TE, and TM modes. LP$_{0\mu}$ modes correspond to $\mathrm{HE}_{1\mu}$ modes: Thus, LP_{01} corresponds to the fundamental HE_{11} mode. $\mathrm{LP}_{1\mu}$ modes are the superposition of $\mathrm{HE}_{2\mu}$, $\mathrm{TE}_{0\mu}$, and $\mathrm{TM}_{0\mu}$ modes, and $\mathrm{LP}_{\nu\mu}$ (with $\nu \geq 2$) modes are the superposition of $\mathrm{HE}_{(\nu+1)\mu}$ and $\mathrm{EH}_{(\nu-1)\mu}$ modes.

The parameter N shown in Figure 7.11 represents a mode number for a planar two-dimensional (2D) waveguide, which is related to the $\mathrm{LP}_{\nu\mu}$ mode in the fiber as

$$N = \nu + 2\mu - 2 \tag{7.3}$$

Since the fiber has a circular three-dimensional structure, the propagating modes are more complex compared to the 2D planar waveguide. However, the LP modes belonging to the same mode number N have almost the same propagation constant, and can be treated as the same principle mode N in the approximate analysis. When $V < 2.405$, only the LP_{01} mode can be propagated, and such a fiber is called a single-mode fiber. When $V > 2.405$, higher order modes can be propagated, and such a fiber is called a multimode fiber. The total number of modes N_T propagating in the step index fiber is approximately given by

$$N_T = V^2/2 \tag{7.4}$$

for large V.

Bandwidth of the multimode fiber is determined by the difference in group delay among the modes. The difference depends on the refractive index profile. When the refractive index profile of the core varies from a step shape to a graded shape, the bandwidth is increased drastically. This is because group delay differences among modes are smaller for the fiber with the graded shape refractive index profile.

The refractive index profile of the GI fiber is represented by

$$n(r) = n_1[1 - 2\Delta(r/a)^\delta]^{1/2} \tag{7.5}$$

where n_1 is the refractive index at $r = 0$, a is the core radius, and Δ is the relative refractive index difference given by

$$\Delta = \left(n_1^2 - n_2^2\right)/2n_1^2 \tag{7.6}$$

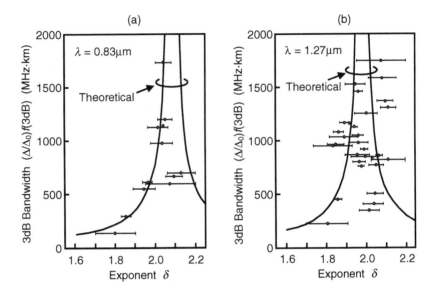

Figure 7.12   Measured and calculated bandwidths as a function of exponent δ for the graded index fiber at (a) 0.83 μm and (b) 1.27 μm.

which is approximated as

$$\Delta \approx (n_1 - n_2)/n_1 \qquad (7.7)$$

for small $(n_1 - n_2)$. Figure 7.12a and Figure 7.12b show measured and calculated normalized 3-dB bandwidths $(\Delta/\Delta_0)f(3 \text{ dB})$ for Ge-doped silica fibers as a function of δ exponent,[11] with $\Delta_0 = 0.01$. Here, Figure 7.12a represents the result at the 0.83-μm wavelength, and Figure 7.12b is that at the 1.27-μm wavelength. It is found that the optimum δ at the 0.83-μm wavelength is 2.08, and that at 1.27 μm is 1.98.

7.5.1.2   Single-Mode Fiber

Contrary to the multimode fiber, bandwidth of the SMF is far broader because it is not relative to the group delay difference among modes. The bandwidth is determined by wavelength dependence of the group delay difference of the fundamental LP_{11} mode in conjunction with a spectral width of the light source. The group delay difference $\Delta\tau$ for a small wavelength difference $\Delta\lambda$ is given by (e.g., Reference 12)

$$\Delta\tau = -L\Delta\lambda[(\lambda/c)(d^2n_1/d\lambda^2) + (N_1\Delta/\lambda c)Vd^2(V\beta_n)/dV^2]$$
$$= \Delta\tau_m + \Delta\tau_g \qquad (7.8)$$

where L is the propagation length; c is the light velocity; and Δ, V, and β_n are given by Equations 7.6, 7.1, and 7.2, respectively. N_1 is the group index and is given by

$$N_1 = n_1[1 - (\lambda/n_1)(dn_1/d\lambda)] \qquad (7.9)$$

According to Equation 7.8, the group delay difference $\Delta\tau$ is given by the sum of two terms: $\Delta\tau_m$ and $\Delta\tau_g$. Here, $\Delta\tau_m$ is given by the first term of Equation 7.8 and originates from the wavelength dependence of the refractive index; $\Delta\tau_g$ is given by the second term and originates from the wavelength dependence of the propagation constant in waveguide structure.

The group delay difference $\Delta\tau$ per unit wavelength difference $\Delta\lambda$ and unit propagation length L is usually defined as the chromatic dispersion D and is represented by

$$D = \Delta\tau/L\Delta\lambda = -(\lambda/c)(d^2n_1/d\lambda^2) - (N_1\Delta/\lambda c)Vd^2(V\beta_n)/dV^2$$
$$= D_m + D_w \qquad (7.10)$$

where D_m is the material dispersion, and D_w is the waveguide dispersion.

Figure 7.13 shows the dispersion curve for the conventional SMF. The total chromatic dispersion is the sum of the material dispersion and the waveguide dispersion, as described in Equation 7.10. Zero dispersion is realized at the 1.31-μm wavelength, and a dispersion value at 1.55 μm is 17 psec/nm/km.

7.5.2 Optical Loss

Optical loss of fiber originates from (1) absorption loss and (2) scattering loss. The absorption losses consist of intrinsic losses of fiber constituent material in the infrared and UV wavelength regions and additional losses because of impurities (e.g., metal ions and OH ion) contained in the fiber. The most important scattering loss originates from Rayleigh scattering. The loss is inversely proportional to λ^4. The other cause of

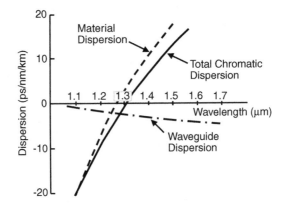

Figure 7.13   Total chromatic dispersion for the conventional single-mode fiber, along with the constituent material dispersion and waveguide dispersion.

the scattering loss is waveguide imperfections occurring in the core and at the boundary between core and cladding.

Figure 7.14 illustrates the total loss, along with the individual loss factors, as a function of wavelength. The dominant loss factors in the visible to infrared wavelength regions in a silica fiber are the Rayleigh scattering and the intrinsic

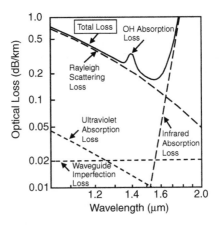

Figure 7.14   Total fiber loss and the individual loss factors as a function of wavelength.

infrared absorption caused by molecular vibrations of the silica glass network. As a result, the lowest loss value of less than 0.2 dB/km is attained at 1.55 µm. Transition metal impurities are thoroughly removed at present by purification of the raw materials, and the OH ion somewhat remains, causing the loss at 1.39 µm. The waveguide imperfection loss averages on the order of 0.01 dB/km.

7.5.3 Standardization of Optical Fibers

7.5.3.1 GI Fiber

At the first stage of optical transmission experiments before 1975, systems mainly utilized step index fibers. Fiber parameters adopted in the United States in that period were typically 110 µm outer diameter and 55 µm core diameter; in Japan, fiber parameters were typically 150 µm outer diameter and 60 µm core diameter. Soon after, GI fibers with a frequency bandwidth wider than several hundreds of megahertz•kilometer, up to about 2 GHz•km, were fabricated successfully owing to improved control techniques for the refractive index profile. Thus, the step index fiber was superseded by the GI fiber by the end of the 1970s. At the beginning of the 1980s, commercial optical transmission systems using the GI fiber were accomplished with the bit rates from several tens to 100 Mb/sec.

Under such situations, efforts to establish a global standard for the GI fiber parameters were started in the International Telegraph and Telephone Consultative Committee (CCITT) (at present ITU-T International Telecommunication Union-Telecommunication Standardization Sector) from 1977. Extensive experimental and theoretical studies were made to determine the optimum fiber parameters.[13,14] Considering a transmission line model, including intrinsic fiber loss, bending loss, connection loss, and so on, relations between the fiber parameters and the total loss for a repeater section of the transmission line model were analyzed. As shown in Figure 7.15, it was found that the total loss was smaller for larger outer diameter $2b$. Choice of the outer diameter was important in view of fiber reliability assurance, fiber cost, and handling ease. Fiber strength was much improved because of

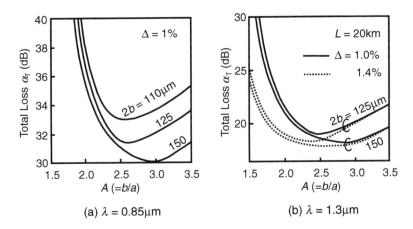

(a) $\lambda = 0.85\mu m$ (b) $\lambda = 1.3\mu m$

Figure 7.15   Total loss of a repeater section of the transmission line model for graded index fibers as a function of ratio A of cladding radius b to core radius a at (a) 0.85 μm and (b) 1.3 μm.

the progress in preform fabrication and drawing processes, and it was concluded that the reliability was fully guaranteed and the total loss was tolerable by choosing an outer diameter of 125 μm. Thus, the optimum ratio A of cladding radius b to core radius a was determined as 2.5 for $2b = 125$ μm to minimize the total loss for the relative refractive index difference Δ of 1.0%. The results were proposed to CCITT, and the proposed parameters (125 μm outer diameter, 50 μm core diameter, and 1.0% relative refractive index difference) were adopted as global standards.

7.5.3.2 Single-Mode Fiber

In parallel with the commercialization of multimode transmission lines, efforts to develop a single-mode transmission line were in progress. Fabrication techniques for the SMF were improved, and fiber with precisely controlled structural parameters and sufficient transmission characteristics was realized early in the 1980s. The optimum SMF parameters were studied to construct a transmission line at a 1.3-μm wavelength. It was found that a combination of a spot size W_0 and an effective cutoff wavelength λ_{ce} was the most suitable compared with the

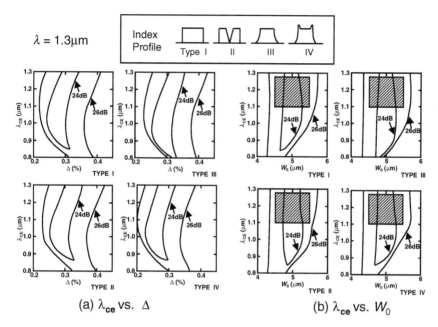

Figure 7.16 Equiloss contours of a repeater section of the transmission line model for various types of refractive index profile for single-mode fibers shown in the inset: (a) contours represented on $\lambda_{ce} - \Delta$ plane and (b) those represented on $\lambda_{ce} - W_0$ plane.

conventional combination of the core radius a and the relative refractive index difference Δ and that of λ_{ce} and Δ.[15]

Figure 7.16 shows the situation. Note that any refractive index profiles are acceptable as well as the ideal rectangular shape profile as far as the single-mode operation is attained. Typical and actually probable profiles in view of fabrication methods are shown in the inset as Types I to IV. Here, Type I is a profile with the ideal rectangular shape, Type II is that with a center dip, Type III is that with a tail, and Type IV is that with a hump and a tail. If the parameters were defined by the combination of a and Δ, equiloss contours were entirely different for each profile type. The situation is somewhat tolerated for the combination of λ_{ce} and Δ, but equiloss contours are still different for each profile, as shown in Figure 7.16a. In the case of λ_{ce} and W_0 shown in Figure 7.16b, the equiloss contours almost coincide for all cases, and the optimum parameter range is represented by a rectangle. Thus, the combination of λ_{ce} and W_0 was the most appropriate to specify the fiber transmission characteristics uniquely for different index profiles. The optimum parameter ranges were determined as W_0 of 5.0 ± 0.5 μm and λ_{ce} of $(1.20 + 0.08/-0.10)$ μm for the SMF for use at 1.3 μm.

The parameter combination to be specified and the concrete parameter values were proposed to CCITT and were accepted as Recommendation G.652 for the global standards.[16] Note that at present the specified parameter values have somewhat varied from the original ones, mainly to attain stronger mode confinement. Representing the presently adopted fiber parameters by $2a$ and Δ for reference, approximate values of $2a$ and Δ are 8 μm and 0.35%, respectively, for a fiber with the step index profile.

SMF transmission systems were introduced around 1983 in trunk networks, with a bit rate of several hundred megabits per second. The systems spread around the world successfully, and coaxial and microwave trunk transmission lines were replaced completely by the fiber lines. At present, a 10-Gb/sec time division multiplexing (TDM) system is used commercially, and a 40-Gb/sec TDM system is under development.

7.5.4 Dispersion-Shifted Fiber

Silica fiber has the lowest loss in the 1.55-μm wavelength region. However, the conventional SMF has a large chromatic dispersion of 17 psec/nm/km in the wavelength region. As demand for high-bit-rate long-distance transmission systems became urgent, it was strongly desired to shift the zero-dispersion wavelength to the 1.55-μm region.

Trials to realize DSF and dispersion-flattened fiber had been made since the mid-1970s, and various refractive index profiles (e.g., W type, triangular type, trapezoidal type, etc.) were proposed (Figure 7.17). A basic idea to realize dispersion-shifted and dispersion-flattened fibers is that a rather complex refractive index profile is adopted, resulting in a distinct change of waveguide dispersion curve. Figure 7.18 schematically illustrates examples of dispersion curves for DSF and dispersion-flattened fiber, along with that for the conventional SMF, for which the zero-dispersion wavelength is 1.31 μm.

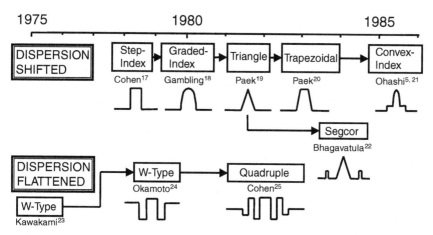

Figure 7.17 Historical flow of research and development for dispersion-shifted fibers and dispersion-flattened fibers.

In 1986, Ohashi et al.[5,21] proposed a novel convex-type profile shown in Figure 7.17. The most serious problems for such types of fiber were bending and microbending loss increases at the wavelength region longer than 1.55 μm. To avoid the loss increases, a 1% total refractive index difference

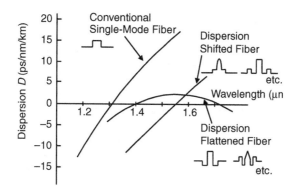

Figure 7.18 Schematic illustration of dispersion curves for the conventional single-mode fiber, a dispersion-shifted fiber, and a dispersion-flattened fiber along with the corresponding refractive index profiles.

was chosen, and the inner core diameter was chosen as approximately 5 μm. The fiber has a zero-dispersion wavelength at around 1.55 μm, a dispersion slope of 0.08 psec/nm²/km, and an effective core area A_{eff} of 40 to 50 μm² at 1.55 μm. The proposed DSF was comparable to the conventional SMF in view of low loss and fabrication ease. Thus, the DSF has been used in the backbone trunk network in Japan, and 2.5- and 10-Gb/sec TDM signals have been transmitted through the fiber. ITU-T standardized[26] the DSF as Recommendation G.653.

7.6 MECHANICAL PROPERTY AND RELIABILITY ASSURANCE OF FIBERS

7.6.1 Fiber Strength

In the earliest developmental stage, fibers were weak and frequently broke far below the empirical maximum tensile failure strength of approximately 5.5 GPa because of the so-called Griffith flaws existing on fiber surfaces. The theoretical breaking stress σ_0 of a material without flaw is given by[27]

$$\sigma_0 = (E\gamma_s/a_0)^{1/2} \qquad (7.11)$$

where E is the Young's modulus, γ_s is the surface energy of the material, and a_0 is the atomic spacing of bond length. For fused silica, $E = 7200$ kg/mm², $\gamma_s = 7 \times 10^{-5}$ kg/mm, and $a_0 = 2 \times 10^{-7}$ mm, and then $\sigma_0 = 16$ GPa (20 kg for a fiber with a 125-μm outer diameter). If a surface flaw with a depth of d_f exists, a breaking stress σ_f given by Griffith[27] is expressed as

$$\sigma_f = (2E\gamma_s/\pi d_f)^{1/2} \qquad (7.12)$$

Combining Equations 7.11 and 7.12,

$$\sigma_f = \sigma_0(2a_0/\pi d_f)^{1/2} \qquad (7.13)$$

When the flaw depth d_f is 1 μm, for instance, σ_f becomes about 1/100 of σ_0.

Figure 7.19 shows typical examples of breaking probability as a function of breaking stress. This figure is called a *Weibull distribution plot*. Solid lines show the failure probability for fibers with a small number of flaws, and the broken

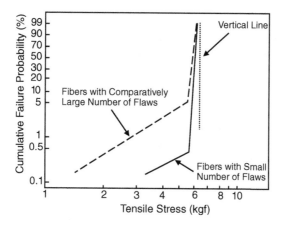

Figure 7.19 Schematic illustration of Weibull distribution plots for cumulative failure probability of fibers with large and small numbers of flaws.

lines indicate fibers with a large number of flaws. It is found from the figure that features at a large breaking stress region are almost the same for both types of fibers. In contrast, features at a small breaking stress region are entirely different, and this results in a decisive difference for fiber reliability in practical transmission lines. The failure probability changes markedly by varying the test fiber length, and use of long test fibers, say 10 m, is desirable.

The fiber strength has been much improved by a number of efforts, such as surface treatment of preform rod, fiber drawing in a clear environment, and appropriate selection of coating materials. As a result, the failure probability is small for short fibers. However, a fiber break still occurs at a small stress value every several to some tens of kilometers, originating from a large-size flaw. Such a failure seriously deteriorates the reliability of long fiber cables under practical circumstances.

Thus, proof testing was proposed to remove large-size flaws before placement. In the test procedure, a predetermined tensile stress is applied continuously along a fiber length for a predetermined period. A break occurs at a flaw that cannot endure the applied stress. Theoretical and experimental investigations were made by Mitsunaga et al.[28] based on the fracture mechanics theory. It was clarified that the failure probability after proof testing could be predicted, even if the initial probability before testing was not known. Thus, fiber reliability was guaranteed according to the applied stress value and the stress-applying period during proof testing.

7.6.2 Bending Loss and Microbending Loss

When a fiber suffers a bend, part of the propagation modes changes to radiation modes, resulting in optical loss. A large loss increase is observed as a bending radius becomes smaller. The loss for each propagation mode depends on its confinement state, and higher modes are apt to radiate at the larger bending radius for the multimode fiber. The threshold value of the radius R at which a large loss increase starts to occur is roughly estimated by[29]

$$R \approx a/\beta_n \Delta \qquad (7.14)$$

where a is the core radius, β_n is the normalized propagation constant given by Equation 7.2, and Δ is the relative refractive index difference between core and cladding given by Equation 7.6. As an example, if a is 5 μm, Δ is 0.003, and β_n is 0.3, then R is about 6 mm.

Small bends occur along a fiber axis when a side pressure is applied to the fiber. Such a bend is called a *microbend*; these cause increased fiber loss. The phenomenon is important for design and manufacturing of optical fiber cables and for a loss increase at low temperature. The microbending loss α_m for a multimode fiber with a plastic coating is given by[30]

$$\alpha_m = (K_m a^4/\Delta^3 b^4 d^2)(E_p/E_f)^2 \qquad (7.15)$$

where K_m is a constant, Δ is the relative refractive index difference, a is the core radius, b is the cladding radius, d is the radius of coated plastic material, and E_f and E_p are the Young's moduli of the fiber and the plastic material, respectively. It was found that the microbending loss depends largely on core, cladding, and coating radii, as well as the refractive index difference. Also, soft plastic material tolerates the loss well.

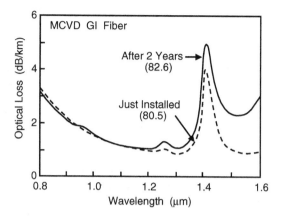

Figure 7.20 Optical loss increase for a GI fiber cable installed in the field. Loss measured just after installation is shown by a broken curve, and that measured after 2 years is shown by a solid curve.

7.6.3 Loss Increase Caused by Hydrogen

In June 1982, a phenomenon was discovered in field-installed optical fiber cables. Fiber loss increased in the wavelength region longer than 1.2 μm (Figure 7.20).[31] The cables were composed of GI fibers and were installed 2 years prior to the loss. The fibers were manufactured by both VAD and MCVD methods and contained various amounts of P_2O_5 and GeO_2 as dopants. The phenomenon was extremely serious and was apt to inflict decisive damage to optical fiber transmission systems. Extensive investigations were started immediately to solve the problem.

A new aspect of loss increase was found in the wavelength region from 1.1 to 1.3 μm for a cable filled with water (Figure 7.21a).[32] It was clarified that the loss peaks coincided completely with those of H_2 molecules.[33] Furthermore, it was clarified that the loss increase observed in the field-installed cables was caused by formation of Si–OH, Ge–OH, and P–OH bonds by chemical reaction, similar to a laboratory experiment (Figure 7.21b).[34] The process is as follows: Defects are formed in the Si–O network because of the dopants; the permeated H_2 molecules are trapped at the defects; and OH bonds

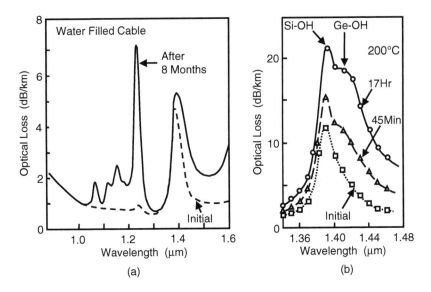

Figure 7.21 Hydrogen-related loss increase phenomena: (a) loss increase caused by H_2 molecule permeated into fiber in a water-filled cable; (b) loss increase caused by formation of OH bonds for a nylon-coated fiber heated at 200°C.

are formed by the chemical reaction. Thus, it was concluded that two types of loss increase existed: One was caused by the H_2 molecule permeated in fiber (Figure 7.21a), and the other was caused by the OH bond formed by the chemical reaction (Figure 7.21b). The latter OH loss is irreversible; the former H_2 loss is reversible as far as H_2 does not react with O_2.

Fundamental countermeasures to prevent the loss increases are that (1) the defect formation in the Si–O network should be minimized and (2) the generation of H_2 should be reduced as much as possible. The former countermeasure is realized to a considerable extent using GeO_2 as a dopant instead of codopants of GeO_2 and P_2O_5. As to the latter countermeasure, H_2 essentially originates from the plastic coating of the fiber, and H_2 generation is prevented by complete polymerization of the coating materials during the fiber-drawing process. Another origin, such as electrolytic corrosion of constituent metals, is probable in a water-penetrated fiber cable. Although the complete prevention of loss increases caused by hydrogen cannot be made, it is possible to suppress them to a minimum, and the fiber reliability is ensured sufficiently under actual practical circumstances for a period longer than 20 years.[35]

7.7 REQUIREMENTS FOR FIBERS IN NEW-ERA TRANSMISSION SYSTEMS

7.7.1 Optical Fiber Amplifier and WDM Transmission System

In 1989, a practical optical fiber amplifier pumped by a LD was reported by Nakazawa et al.[6]; immediately, a high-bit-rate long-distance transmission system using the fiber amplifier was reported by Hagimoto et al.[36] Realization of the practical Er-doped fiber amplifier changed transmission system aspects drastically. The most important progress is that long-distance dense WDM transmission systems are practical because of the simplified system configuration and reduced system cost. Rapid increase in data traffic because of the explosive growth of the Internet triggered the introduction of the dense WDM systems in commercial transmission links mainly in the United States. The number of wavelength channels, which starting from 4 to 8, increased to 32 to 64 in the commercial systems, and the possible wavelength number has increased to 1000 on the laboratory level.[37]

The new revolution of optical transmission systems has strongly influenced specifications for fiber transmission characteristics. First, influence of the fiber chromatic dispersion was severe because, by using the optical amplifier as a repeater, data signals are propagated at a longer distance without optical/electrical (O/E) and electrical/optical (E/O) conversions and because a wider wavelength region is required for the large-capacity WDM system. Second, optical nonlinear effects become severe because the WDM increases the power of the total transmitted signals and narrow wavelength spacing causes interference between adjacent wavelength signals.

Below, influences of the nonlinear effects and the chromatic dispersion on fiber transmission characteristics are discussed.

7.7.2 Optical Nonlinearity in Fibers

Optical nonlinear effects in fiber are not essentially large and have been taken into consideration only slightly regarding fiber transmission characteristics over a long duration. However, because optical fiber amplifier and WDM have been introduced in commercial systems and the transmission bit rate increased, the nonlinear effects play an important role, good or bad, in transmission system performance.

The optical nonlinear effects in fiber are summarized in Table 7.2. The magnitude of nonlinearity in fiber depends on the (1) nonlinear coefficient, (2) optical power input, (3) effective area A_{eff} of the optical mode field related to the field distribution (4) effective fiber length related to the optical fiber loss, and (5) phase-matching condition related directly to the fiber dispersion. Among various nonlinear effects, four-wave mixing (FWM) drastically affects transmission signal quality for the dense WDM system. Figure 7.22 illustrates schematically the generation of FWM waves. By the combination of three waves with frequencies f_i, f_j, and f_k, a new wave with a frequency of f_{ijk}, given by

$$f_{ijk} = f_i + f_j - f_k \qquad (7.16)$$

TABLE 7.2   Optical Nonlinear Effects in Fiber

Origin of nonlinearity	One Wavelength (Single channel)	Multiwavelength (Multichannel)
Scattering	Stimulated Brillouin scattering (SBS)	Stimulated Raman scattering (SRS)
Third-order susceptibility	Self-phase modulation (SPM)	Cross-phase modulation (XPM) Four-wave mixing (FWM)

Magnitude of nonlinearity depends on

Nonlinear coefficient		
Optical power input		
Effective area distribution	$\Leftarrow$	Optical mode field distribution
Effective fiber length	$\Leftarrow$	Optical loss
Phase-matching condition	$\Leftarrow$	Dispersion

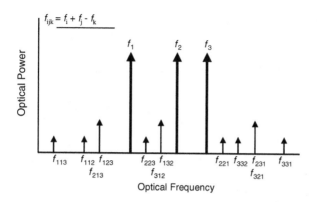

Figure 7.22   Generation of FWM waves by the combination of three waves.

is generated. Figure 7.23 shows the power of the FWM wave as a function of wavelength spacing of WDM signals.[38] The fiber dispersion is taken as a parameter. It is shown from Figure 7.23 that the fiber dispersion strongly affects the generation of the FWM wave, and wider wavelength spacing is required to prevent it for a smaller dispersion value.

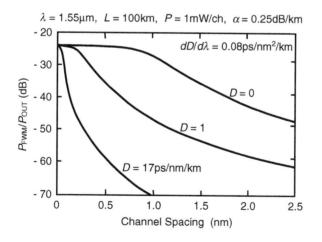

Figure 7.23   Generated power of the FWM wave as a function of WDM channel spacing with chromatic dispersion D as a parameter.

Examples of positive use of the nonlinear effects are soliton transmission,[39] supercontinuum light source,[40] and Raman amplification.[41] A principle of soliton transmission is that the self-phase modulation (SPM) and the fiber dispersion are compensated in an anomalous dispersion region, and a transmitted pulse width is maintained during transmission.[39] In an actual fiber, intensity of the optical pulse decreases along a fiber length because of the fiber loss, and the ideal soliton transmission cannot be realized. However, by repeating the optical amplifications for the weakened optical power, effective soliton transmission is realized.

The supercontinuum light source[40] is composed of a mode-locked fiber ring laser, a high-power optical amplifier, and a dispersion-managed special fiber, for example, with wavelength dependence of the dispersion that is convex and the dispersion value decreases from positive to negative along the fiber length. A 200-nm wide, low-noise, continuous spectrum is obtained and is expected as a light source for an ultradense and ultrawide WDM system and as a wavelength-variable, short-pulse generator.

The Raman amplifier[41] has essentially no wavelength limitation and is fully expected for amplification in the wide-wavelength region, in which no adequate rare-earth-doped fiber amplifier is available. Especially, a rare-earth-doped amplifier associated with Raman amplification is effective in flattening the amplifier gain and reducing the noise. It is advantageous that the main transmission fiber is used as an amplification medium and that the amplifier bandwidth can be widened by simultaneous use of a number of pump lasers with different wavelengths. As an example, a 100-nm band flat amplification with 6-dB net gain was reported.[42] A dispersion-compensating fiber (DCF), which is described in Section 7.8.2, is also useful as a Raman amplification medium because the DCF has a small A_{eff} and a large nonlinear coefficient because of the considerable amount of dopant.

7.7.3 Chromatic Dispersion

The transmission capacity and distance are limited by the fiber chromatic dispersion, as described in Section 7.5.1.2.

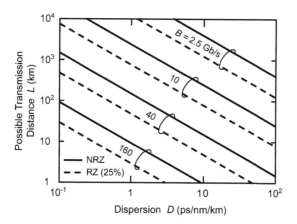

Figure 7.24 Possible transmission distance for NRZ and RZ signal formats as a function of chromatic dispersion with the bit rate B as a parameter.

Figure 7.24 shows a possible transmission distance L for an nonreturn-to-zero (NRZ) format and an return-to-zero (RZ) format with a duty factor of 25% as a function of the dispersion D with the transmission bit rate B as a parameter. The chart is drawn based on a basic calculation[43] for which nonlinear effects in fiber and frequency chirping of LD are neglected. The possible transmission distance is inversely proportional to the dispersion value. The allowable dispersion is in inverse proportion to the square of the bit rate because both the optical spectrum broadening and the pulse interval narrowing are proportional to the bit rate.

In the case of 2.5 Gb/sec with a transmission distance around 1000 km, for example, dispersion compensation is not necessary for the NRZ signal even for the conventional SMF with a 17-psec/nm/km dispersion at a 1.55-μm wavelength insofar as the frequency chirping of the LD is small. For 10-Gb/sec transmission, the situation changes entirely, and an allowable dispersion value is restricted to the order of 1 psec/nm/km for a transmission of the same distance. To compensate for the large dispersion and to make the high-bit-rate and large-capacity transmission possible, a DCF is necessary. This is discussed in Section 7.8.2.

It should be noted that the frequency chirping of a directly modulated Fabry–Perot LD is fairly large and may affect the transmission system performance considerably. Therefore, evaluation of the actual system performance must be made, taking into account the fiber chromatic dispersion, fiber nonlinear effects, and the LD spectrum broadening, including the chirping effect.

7.7.4 Polarization Mode Dispersion

Figure 7.25 is a schematic illustration of polarization mode dispersion (PMD). If birefringence occurs in a fiber because of stress, core ellipticity, and so on, two degenerating orthogonal modes appear. The modes are propagated with different speeds, and this causes the PMD. The PMD effect is serious for the high-bit-rate, long-distance transmission. Figure 7.26 shows a possible transmission distance as a function of PMD D_{PMD}, with the transmission bit rate as a parameter. Here, the following equation derived by Kapron[44] is used:

$$L \text{ (km)} = [10^3 f / \{B(\text{Gb/sec}) \cdot D_{\mathrm{PMD}}(\text{psec/km}^{1/2})\}]^2 \quad (7.17)$$

A parameter f is given by

$$f = PMD \cdot B, \quad (7.18)$$

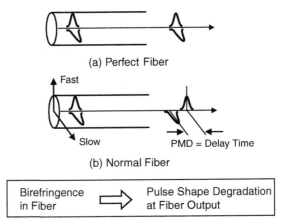

(a) Perfect Fiber

Fast

Slow   PMD = Delay Time

(b) Normal Fiber

| Birefringence in Fiber | ⟹ | Pulse Shape Degradation at Fiber Output |

Figure 7.25 Schematic illustration of the PMD in a birefringent fiber.

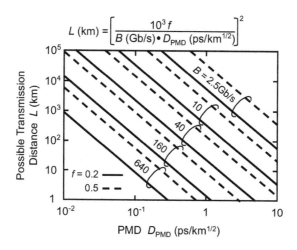

Figure 7.26 Possible transmission distance as a function of PMD with the bit rate B as a parameter.

where *PMD* is the total PMD, and B is the bit rate. The f value of 0.2 is regarded as a proper value considering statistical features of the time-varying PMD, and the value of 0.5 is the most optimistic one without taking into account the statistical nature of the PMD fluctuation. The root of L is in inverse proportion to the PMD value. This is because mode coupling occurs between the orthogonal modes at random during transmission, and this widens the tolerance of the transmission distance favorably. The PMD of field-installed fibers varies incessantly with time in a complex manner, and an automatic PMD compensator that follows up the real-time variation well is necessary to prevent degradation of transmission signal quality.

7.8 NOVEL FIBER FOR THE WAVELENGTH DIVISION MULTIPLEXING SYSTEM

7.8.1 Fibers for Main Transmission Lines

Since the WDM transmission systems accompanied by optical amplifiers were introduced and higher bit rate transmission was realized, it became necessary to develop novel fibers suitable for such systems and devices compensating the fiber chromatic dispersion in wide-wavelength regions. Novel fibers developed for the requirements have included TrueWave fiber,[45] TeraLight® (Alcatel, France) fiber,[46] and LEAF.[47]

A basic idea of the fibers is that the dispersion at the C-band (1.53 to 1.565 µm), which is the main amplification region of the Er-doped fiber amplifier, is small but not zero to prevent nonlinear effects appearing strongly because of phase matching. As for LEAF, an effective area A_{eff} of the mode field is also enlarged for further reduction of the FWM wave generation. Various types of fibers have been developed: fibers with zero-dispersion wavelengths shifted to longer and shorter sides than the C-band and fibers with reduced dispersion slopes. ITU-T standardized[48] these nonzero dispersion-shifted fibers (NZ-DSFs) as Recommendation G.655. The conventional DSF with zero dispersion around 1.55 µm is standardized[26] as G.653. Table 7.3 summarizes typical examples of the developed NZ-DSFs. Here, values of A_{eff}, dispersion D, dispersion slope S, and fiber loss α at the 1.55-µm wavelength are listed, along with a profile type of the refractive index. The fibers listed in the table include large effective area fibers with A_{eff} larger than 70 µm² and slope-reduced fibers with S smaller than 0.04 psec/nm²/km.

Figure 7.27 shows an example of the dispersion curve for dispersion-flattened fibers.[55] The fiber exhibits a dispersion slope of zero at 1.55 µm and a dispersion deviation within 1 psec/nm/km in the whole wavelength region covering the S-band (1.46 to 1.53 µm), C-band, and L-band (1.565 to 1.625 µm). An extremely precise control of fiber parameters is required to obtain dispersion-flattened fiber with the designed values of dispersion and dispersion slope. However, the fiber is considered useful for future ultrawideband WDM systems, and further improvement of fabrication techniques is strongly desired.

High-bit-rate WDM transmission with a narrow wavelength spacing, that is, with high spectral efficiency (bits/second/hertz) is now required. For such a system, it is important to reduce the nonlinear effects as much as possible and to suppress a cumulative dispersion value at the middle of the transmission line to be moderate. Thus, a novel NZ-DSF called a medial dispersion fiber (MDF) with a dispersion value of around 10 psec/nm/km is proposed.[52,56]

TABLE 7.3 Refractive Index Profiles and Performances of Typical Nonzero Dispersion-Shifted Fibers

Type	Profile	A_{eff} (μm^2)	D (psec/nm/km)	S (psec/nm^2/km)	α (dB/km)	Ref.
Convex		56	−1.7	0.094	0.204	49
Segment		56	3.7	0.046	0.204	45
		83–98	−3.0~+1.9	0.10–0.11	0.21–0.24	50
Segment + Trench		55–72	−4~−0.5 +0.5~+6	0.014–0.04	0.23–0.45	47
Ring		>90	−2~+2	0.086	0.23	51
		104	9.2	0.08	0.23	52
Double rings		55–120	−3~+3	0.07–0.1	—	53
W + ring		64, 67	2.2, 2.7	0.032, 0.035	0.235, 0.21	54
		46	5.2	0.000	0.211	55

Note: Values at 1.55 μm are listed.

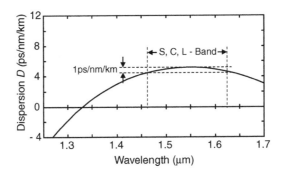

Figure 7.27 An example of a dispersion-flattened fiber. Dispersion slope value of complete zero is attained at 1.55 μm, and dispersion deviation is within 1 psec/nm/km in the whole S-, C-, and L-bands.

As far as the C-band is concerned, existing transmission lines consisting of the DSFs with zero dispersion around 1.55 μm are disadvantageous for WDM systems. A wide and unequal wavelength spacing must be adopted to prevent the influence of FWM. However, the use of L- and S-bands becomes ordinarily possible by using gain-shifted amplifiers and Raman amplification, and the disadvantage at the C-band is mitigated to a tolerable extent. When a wider wavelength band in one fiber is requested in the future for an ultralarge-capacity transmission system, any NZ-DSF except that with a sufficiently small dispersion slope will suffer the same drawback in its zero-dispersion region.

7.8.2 DCF with Large Dispersion Value

As described in Section 7.7.3, the chromatic dispersion of fiber limits the bit rate and the transmission distance. In the conventional transmission system, in which O/E and E/O conversions are made at every repeater, pulse signals broadened by the dispersion are reshaped, and completely well-shaped signals are regenerated. Therefore, the total desired transmission distance through the repeaters has no limit as far as the transmission within one repeater spacing, namely, several tens to 100 kilometers, is possible. While in the system using optical amplifiers as the repeaters, transmitted signals are

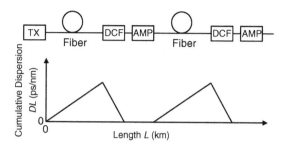

Figure 7.28   Transmission line model and cumulative dispersion value as a function of the propagation distance. AMP, amplifier; DCF, dispersion compensating fiber; TX, transmitter.

only amplified at the repeater with almost no change in pulse shape. Therefore, it must first be checked whether pulse signals can be recognized after they transmit at the desired distance through a number of amplifiers. If the signals cannot be recognized, adequate dispersion compensation is required.

In principle, total dispersion of the whole length of transmission lines is given simply by the sum of dispersion values of individual constituent fibers. Thus, dispersion compensation is possible by connecting a fiber with an adequate dispersion value. Figure 7.28 illustrates a typical transmission system using DCFs and optical amplifiers. Here, it is assumed that the dispersion of main transmission fibers is positive. Dispersion increases along the fiber length in the main transmission fiber and decreases in the DCF, resulting in realization of the total zero dispersion at the fiber amplifier.

Dispersion compensation first began for high-bit-rate systems using one wavelength around 1.55 μm through a conventional SMF with a dispersion of 17 psec/nm/km.[57] Since that time, various practical DCFs have been developed with a large negative dispersion of –50 to –100 psec/nm/km at 1.55 μm with a positive dispersion slope.[58] Thus, compensation was achieved at one desired wavelength by using a DCF with a length that was one third to one sixth of transmission fiber. This type of DCF applicable to one wavelength may be called the first-generation DCF. The DCF has a large refractive index difference and is apt to exhibit a rather high loss value. To evaluate the compensation ability, the figure of merit (FOM) M is defined as[58]

$$M = D/\alpha(\text{psec/nm/dB}) \qquad (7.19)$$

where D (psec/nm/km) is the dispersion, and α (dB/km) is the fiber loss. Typically, an M around 300 psec/nm/dB has been attained.

Since WDM systems became common commercially, an advanced type of DCF has become necessary, one with functions to compensate the dispersion slope together with the dispersion value to achieve compensation in the wide-wavelength region. This may be called the second-generation DCF. Such a fiber should have a large negative dispersion slope S as well as a large negative dispersion value D. These characteristics are realized with a depressed cladding-type refractive index profile.[59] Table 7.4 summarizes typical values of A_{eff}, D, S, and α at the 1.55-µm wavelength for the developed DCFs. The first generation DCF with a large negative value of D and a positive slope S is shown in the first line for reference. This table also lists a ratio D/S, dispersion per slope (DPS), that is, reciprocal of relative dispersion slope (RDS), and the figure of merit M. It is beneficial to use D/S, instead of RDS, because of the adequate figure size for D/S. The value of D/S for the conventional SMF is approximately 300, and a DCF with D/S close to 300 suits the fiber for wideband compensation. For high-bit-rate, large-capacity WDM systems, even nonzero DSFs require compensation, and DCFs compatible with the nonzero DSFs also have been developed.[62,67] The value of D/S ranges typically from some tens to 100.

7.8.3 Reverse Dispersion Fiber

The DCF described in Section 7.8.2 is wound tightly on a reel and is installed in transmission equipment as a module. The loss of the module is added to the transmission line as the excess loss. A new idea was proposed that a fiber for compensation be deployed in a cable and be connected with a main transmission fiber, constituting a part of the actual transmission line. The line thus constructed is called a *dispersion-managed*

TABLE 7.4 Refractive Index Profiles and Performances of Typical DCFs (for Both Dispersion and Dispersion Slope Compensation)

Profile	A_{eff} (μm²)	D (psec/nm/km)	S (psec/nm²/km)	α (dB/km)	D/S (nm)	M (psec/nm/dB)	Ref.
	17	−98	+0.1	0.35	—	280	60
W – type	15	−134	−0.2	0.67	670	200	60
	—	−133	−0.55	0.69	240	190	61
	—	−25,−7	−1.8,−1.6	—	14,4	—	62
W + ring	19	−168	−1.59	0.75	105	225	63
	19	−232	−2.24	0.59	104	395	64
	—	−105	−0.33	0.54	320	195	65
	21	−62	−0.21	0.38	290	160	66

Note: The first line shows the first-generation DCF compensating only the dispersion at one wavelength. Values at 1.55 μm are listed.

line. An example was a proposal made by Mukasa et al.,[68] and the fiber for compensation was designated a reverse dispersion fiber (RDF® [Furukawa Electric, Japan]). By connecting the RDF with the conventional SMF, a 127-km transmission line was realized in the 50-nm wavelength bandwidth.[69] This type of transmission line is considered especially useful for submarine systems.

Table 7.5 summarizes performance characteristics of a set of developed RDFs.[67,70] The fiber developed by Grüner-Nielsen et al.[67] is represented as an inverse dispersion fiber (IDF). The RDF of the first stage was prepared so that the length ratio of RDF to SMF was 1:1. However, the RDF has a rather small A_{eff}, and the transmission performance possibly is affected by the FWM wave. There were trials to shorten the RDF length to a tolerable extent, up to 1:3, by increasing the dispersion and slope values with D/S invariable.

Optical loss of the RDF generally is larger than that of the main fiber to be compensated, and it is necessary to evaluate the loss penalty because of the compensation. Loss increment $\Delta \alpha_T$ of the total transmission line is given by

$$\Delta \alpha_T = L_R(\alpha_R - \alpha_M)$$
$$= D_M L_T(\alpha_R - \alpha_M)/(D_M - D_R) \qquad (7.20)$$

where the subscripts R and M denote the RDF and the main fiber, respectively, and L_T is the total transmission line length. Thus, a new FOM applicable to the RDF can be defined as

$$M_R = (D_M - D_R)/(\alpha_R - \alpha_M) \text{ (psec/nm/dB)} \qquad (7.21)$$

Values of M_R listed in the table were calculated using a D_M of 17 psec/nm/km and an α_M of 0.19 dB/km for the conventional SMF. The loss penalty is smaller for higher M_R.

The idea of RDF was also applied to the compensation of NZ-DSF. It was demonstrated that dispersion and dispersion slope of TeraLight fiber were compensated by the combination of TeraLight/reverse TeraLight with a 2/1 length ratio.[71] Combination of an MDF, described in Section 7.8.1, and a pair of another MDF with the reverse dispersion and slope values was also proposed.[72]

TABLE 7.5 Performances of Reverse Dispersion Fiber (RDF) and Inverse Dispersion Fiber (IDF) for Compensating the Conventional Single-Mode Fiber with 1:1 to 1:3 Length Ratios

Type		A_{eff} (μm^2)	D (psec/nm/km)	S (psec/nm²/km)	α (dB/km)	D/S (nm)	M_R^a (psec/nm/dB)	Ref.
R	~1:1	24	−24	−0.060	0.24	400	820	
D	~1:2	21	−40	−0.130	0.26	310	815	70
F	~1:3	23	−60	−0.228	0.255	260	1180	
I	~1:1	35	−17	−0.054	0.23	315	850	
D	~1:2	30	−40	−0.12	0.26	330	815	67
F	~1:3	26	−54	−0.16	0.29	340	710	

[a] $M_R = (D_M - D_R)/(\alpha_R - \alpha_M)$. D_M and α_M are chosen as 17 psec/nm/km and 0.19 dB/km, respectively, for the conventional single-mode fiber.

Note: Values at 1.55 μm are listed.

7.8.4 Limitation of WDM Transmission Ability Because of Fiber Dispersion

As described in Section 7.7.3, an allowable fiber chromatic dispersion is inversely proportional to the square of the transmission bit rate. This is because both optical spectrum broadening and pulse interval narrowing are proportional to the bit rate. Severe limitations appear for the WDM transmission system with a bit rate higher than 10 Gb/sec, and strict dispersion slope compensation is required.

Figure 7.29 is a schematic illustration of dispersion compensation for the WDM transmission. Compensation must take place throughout the wavelength region from ($\lambda_0 - \Delta\lambda/2$) to ($\lambda_0 + \Delta\lambda/2$) by using a DCF with appropriate dispersion and slope values. Parameters of the transmission fiber and the DCF are shown in Figure 7.29. The value of L_T/L_C is usually several. For simplicity, it is assumed that the dispersion changes linearly with wavelength; that is, third-order dispersion is neglected for both fibers. It is assumed that the dispersion is completely compensated at λ_0 and is not compensated at other

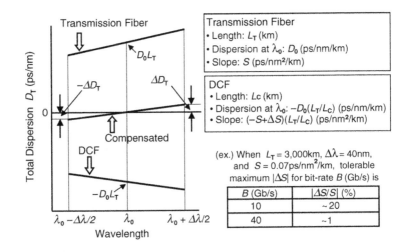

Figure 7.29 Schematic illustration of dispersion and dispersion slope compensation. Tolerable slope deviations ΔS against the transmission bit rate are also shown. DCF, dispersion-compensating fiber.

wavelengths, depending on the slope deviation $\Delta S(L_T/L_C)$ of the DCF. The deviation ΔD_T at $(\lambda_0 + \Delta\lambda/2)$ is given by

$$\Delta D_T = \Delta S \Delta\lambda L_T/2 \qquad (7.22)$$

A table inserted in Figure 7.29 shows a tolerable slope deviation $|\Delta S|/S$ of the DCF for 10- and 40-Gb/sec transmissions when $L_T = 3000$ km, $\Delta\lambda = 40$ nm, $S = 0.07$ psec/nm^2/km, and RZ format with a duty factor of 25%. The compensation is made generally at every repeater section, and ΔS is an average deviation of a number of DCFs usually located in front of the amplifiers. It seems rather easy to fabricate and prepare the DCF for 10-Gb/sec transmission because the slope deviation of approximately 20% is allowed. In contrast, it suddenly becomes difficult to prepare the DCF for 40-Gb/sec transmission: Only approximately 1% deviation is allowed on average for a number of DCFs. As far as the present state of the art is concerned, large-capacity, long-distance WDM transmission with a bit rate higher than 40 Gb/sec is not advisable because of the heavy burden on the DCF.

Note that the total transmission capacity is invariable for different systems, insofar as the same wavelength bandwidth is occupied and the same spectral efficiency (i.e., the bit rate/frequency ratio [bits/second/hertz]) is adopted, regardless of the bit rate. Therefore, an increase in bit rate using a means other than purely electrical TDM is not necessarily advantageous: With electrical TDM, there is an increased possibility of reducing the costs of transmitter and receiver equipment. To overcome the bit rate limitation mentioned, it is indispensable to realize an optical 3R repeater, which may compensate the residual pulse broadening beyond the DCF's ability. It is also useful to realize inexpensive, small, versatile compensation devices other than fiber. Use of DCFs associated with such devices may be effective.

Figure 7.30 illustrates schematically a desirable combination of the bit rate per channel B and the channel spacing on the basis of the present and near-future states of the art. In this figure, total capacity is denoted as B_T. Design and fabrication techniques for WDM filters have improved, and wavelength spacing narrower than 0.1 nm is possible.[37] Furthermore,

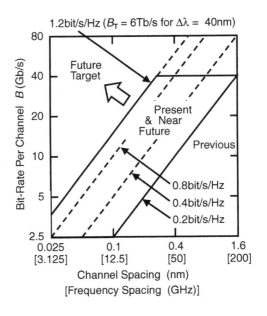

Figure 7.30 Desirable combination of the bit rate per channel and the channel spacing in view of the present and near-future states of the art.

spectral efficiency (bits/second/hertz) exceeding unity is expected by applying efficient modulation formats and filtering techniques, such as *M*-ary transmission coding format,[73] vestigial sideband modulation and filtering,[74] and optical code division multiplexing (OCDM),[75] and by using an interleaver[76] associated with the WDM filter.

7.8.5   Summary of Fiber Development Flow

Figure 7.31 summarizes an outline of fiber development flow, starting with the conventional SMF. Requirement of higher bit rate transmission using the lowest loss region (i) caused the development of a convex index-type DSF.[5,21] Appearance of an optical fiber amplifier[6] and the requirement for long-distance transmission using the conventional SMF mainly made in the United States (ii) triggered the development of the first-generation DCF[57] applicable to one wavelength. Longer distance transmission of higher bit rate

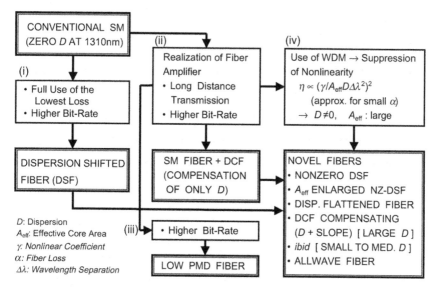

Figure 7.31 Flow of fiber development stimulated by various requirements described by items i to iv. DCF, dispersion-compensating fiber; Disp., dispersion; NZ-DSF, nonzero dispersion-shifted fiber; PMD, polarization mode dispersion; SM, single mode.

signals (iii) led the reduction efforts for the PMD.[44] Introduction of WDM transmission systems realized by the fiber amplifier caused severe fiber nonlinearity (iv), and this resulted in the development of NZ-DSFs[45,46], fibers with large effective core area[67], the second-generation DCF,[59] and so on. Thus, at present.total transmission capacity of several hundreds of gigabits/second in a fiber is realized commercially. An ultimate large capacity of 10 Tb/sec has been demonstrated in laboratory experiments.[74,77]

7.9 FIBER FOR ACCESS, METRO, AND LOCAL-AREA NETWORKS

The conventional SMF has a simple refractive index profile and is attractive in view of fabrication ease and superior low-loss characteristics. Therefore, use of conventional SMF is effective for a system utilizing the 1.3-μm wavelength and for a system utilizing 1.55 μm as long as the fiber dispersion at the wavelength region does not limit the system performance. In Japan, extensive introduction of fibers in an access network, fiber to the home (FTTH), is in progress. The access transmission lines are composed of conventional SMF, and the transmission signals of both 1.3- and 1.5-μm wavelength regions are utilized.

A metro network covers areas up to several hundred kilometers long. The plural number of metro networks hang down on the long-haul backbone network. Among the metro networks, the metro–core network covers a distance from about 60 to 600 km, and the metro–access network, which is located near the subscriber side, covers up to about 60 km. As for the metro–core network, dense WDM transmission systems with bit rates of 2.5 and 10 Gb/sec are utilized. The systems are composed of NZ-DSFs with optical fiber amplifiers and DCF modules. The configuration is similar to that of the backbone network but is simpler and less expensive, taking into account the decreased system requirements. As for the metro–access network, a coarse WDM system with a 2.5-G/sec bit rate is utilized. The configuration of the metro–access network differs considerably from that of the metro–core network, and it is possible to use conventional SMF instead of the NZ-DSF in most cases.

Conventional SMF has ordinarily been used in the 1.55-μm wavelength region. However, in some cases there has been serious bending and microbending loss increase. To avoid this problem, a new fiber has been developed with a cutoff wavelength that is shifted to around the 1.5-μm wavelength region by increasing the core diameter or the refractive index difference between core and cladding. ITU-T standardized[78] this cutoff shifted fiber as Recommendation G.654. The bending and microbending losses are reduced drastically by several orders. The dispersion at 1.55 μm is somewhat increased up to 20 psec/nm/km at the maximum.

An optical local-area network (LAN) covers various application areas and exhibits various system constructions in accordance with the system object and transmission capacity and distance scale. In most cases, the GI fiber or conventional

SMF will be the best choice, depending on the system requirements. Full use of the wider wavelength region, say 1.2 to 1.7 µm, for which the fiber loss is below 0.5 dB/km, is attractive, and AllWave® (OFS-Fitel, Inc., U.S.) fiber has been developed for such purposes. For the further development of the AllWave fiber, it is necessary to establish techniques for complete reduction of the OH ion in the fabrication processes and reliability assurance for a long period.[79] In the future, utilization of a number of wavelengths, say more than a few thousand, will be effective, especially for the LAN system, which has a large scale but a moderate transmission distance. In such a system, it is possible to allocate one wavelength, for example, to each terminal.

Specifications for 10-Gb/sec Ethernet have been standardized as IEEE802.3ae. Table 7.6 summarizes the fiber specifications used for the 10-Gb/sec Ethernet.[80] The GI fiber with a 50-µm core diameter (designated 50MMF), that with the 62.5-µm core diameter (62.5MMF), and conventional SMF can be chosen according to system requirements. As for the SMF, fibers other than the conventional one, such as a NZ-DSF, can be used.

7.10 FUTURE PROSPECTS FOR OPTICAL FIBERS

7.10.1 Fibers for Transmission Lines

Finally, I discuss future prospects for optical fibers in anticipation of a photonic network. As a total transmission capacity in one fiber becomes larger by the increased bit rate of one channel or the increased channel number in the WDM systems, requirements of fiber performance are more severe. Thus, further improvement in fabrication techniques is indispensable for main transmission fibers and DCFs. The techniques include (1) precise control of fiber structural parameters, dispersion value, and dispersion slope; (2) excellent uniformity along a fiber's length; and (3) high manufacturing repeatability. Reduction of PMD is also an important item. Although the development of dynamic PMD compensating devices is indispensable, PMD of fiber itself should be diminished to the 0.01-psec/km$^{1/2}$ level. Reduction of the dispersion slope (ideally, the realization

TABLE 7.6 Fiber Specifications for the 10-Gb/sec Ethernet

Fiber (MHz · km)	Coding	62.5 MMF		50 MMF			SMF
		160	200	400	500	2000	
SR/SW 850 nm	64B/66B	26 m	33 m	66 m	82 m	300 m	—
LR/LW 1310 nm	64B/66B	—	—	—	—	—	10 km
ER/EW 1550 nm	64B/66B	—	—	—	—	—	30–40 km
LX4/LW4 1310 nm[a]	8B/10B	300 m at 500 MHz · km		240 m	300 m	—	10 km

[a] Use of coarse wavelength division multiplexing (CWDM) (4 wavelengths, approximately 20-nm spacing), 3/.125 Gb/sec. × 4.

Note: MMF, multimode GI fiber; SMF, single-mode fiber.

of practical dispersion-flattened fibers) along with development of the corresponding DCFs are important for wideband WDM systems.

As for rare-earth-doped fibers for fiber amplifiers, preparation of amplifier menus throughout the usable wavelength region is desirable. The commercial introduction of a Raman amplifier has begun. Also, improved efficiency of a fluoride fiber, especially Tm-doped fiber,[77] is noteworthy. After these fruitful results, further development of various types of amplifiers and their constituent fibers and devices is desired.

7.10.2 Fiber-Type Devices

There exist many important and attractive research items for fiber-type devices. One is the extension of application areas for optical nonlinear effects in fiber, and another is the improvement of photonic crystal fibers already developed.[81]

Table 7.7 lists two types of the photonic crystal fibers. The first is called a holey fiber, and the second is called a photonic band-gap fiber. Principles of optical waveguiding are as follows: For the holey fiber, optical wave confinement is realized by the

TABLE 7.7 Structures and Characteristic Features of Photonic Crystal Fibers

Fiber type	Waveguiding	Characteristic features
Holey Fiber	Effective refractive index difference between core and cladding	Wide range of dispersion value and slope Wide range of effective area Easily conferred birefringence Single-mode propagation in ultrawide wavelength region
PBG Fiber	Confinement by photonic band gap associated with implanted defect	Free selection of core material Vacuum core, air core may yield ultralow loss ultrasmall dispersion ultralow nonlinearity

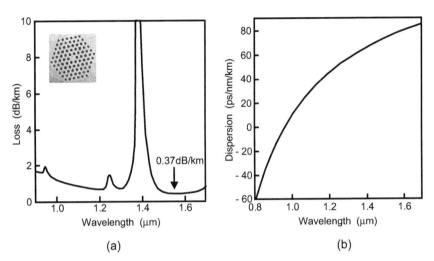

Figure 7.32   Examples of (a) loss spectrum and (b) dispersion for a holey fiber.

effective difference of refractive indices between core and cladding regions, similar to the conventional fiber. For the photonic band-gap fiber, optical wave confinement is realized because of the principle of the photonic band-gap associated with an implanted defect, which forms a core.

These fibers have several unique characteristic features (Table 7.7) that conventional fibers can never exhibit. Thus, various useful fiber-type devices with novel performance and structures are realizable. Figure 7.32 shows transmission characteristics of a recently developed holey fiber.[82] An optical loss as low as 0.37 dB/km was realized successfully. This low loss value may become a milestone for future advancement of the photonic crystal fibers.

Developments of various functional devices for signal processing, such as wavelength converter, pulse compressor, noise filter, clock extraction device, and frequency up- and down-converters, are expected. Optical fibers, as well as LDs and semiconductor optical amplifiers (SOAs), will play an important role in the realization of the devices for signal processing. The challenging target is an optical 3R repeater. If the optical 3R repeater with sufficient performance is realized, it can act exactly the same as the conventional repeater with O/E and E/O conversions. Thus, the transmission systems with a bit rate higher than 100 Gb/sec, irrespective of electrical TDM or optical TDM, may be realized comparatively easily. The functional devices mentioned here will also play an important role in future photonic Internet protocol (IP) routers used in the photonic network.

REFERENCES

1. K.C. Kao and G.A. Hockham, Dielectric-fibre surface waveguides for optical frequencies, *Proc. IEE*, 113, 1151 (1966).

2. F.P. Kapron, D.B. Keck, and R.D. Maurer, Radiation losses in glass optical waveguides, *Appl. Phys. Lett.*, 17, 423 (1970).

3. T. Miya, Y. Terunuma, T. Hosaka, and T. Miyashita, Ultimate low-loss single-mode fibre at 1.55 μm, *Electron. Lett.*, 15, 106 (1979).

4. K. Nagayama et al., Ultra low loss (0.151 dB/km) fiber and its impact on submarine transmission systems, *OFC*, PD Paper, FA10 (2002).

5. M. Ohashi, N. Kuwaki, C. Tanaka, N. Uesugi, and Y. Negishi, Bend-optimised dispersion-shifted step-shaped-index (SSI) fibres, *Electron. Lett.*, 22, 1285 (1986).

6. M. Nakazawa, Y. Kimura, and K. Suzuki, Efficient Er^{3+}-doped optical fiber amplifier pumped by a 1.48 μm InGaAsP laser diode, *Appl. Phys. Lett.*, 54, 295 (1989).

7. J.B. MacChesney, P.B. O'Connor, F.V. DiMarcello, J.R. Simpson, and P.D. Lazay, Preparation of low loss optical fibers using simultaneous vapor phase deposition and fusion, *10th Int. Congress Glass*, 6–40 (Kyoto, 1974).

8. D.B. Keck, B. Flats, and P.C. Schultz, Method of Forming Optical Waveguide Fibers, U.S. Patent 3,737,292 (1972).

9. T. Izawa, S. Kobayashi, S. Sudo, and F. Hanawa, Continuous fabrication of high silica fiber preform, *IOOC'77*, C1-1, 375 (1977).

10. D. Gloge, Weakly guiding fibers, *Appl. Opt.*, 10, 2252 (1971).

11. K. Kitayama, S. Seikai, Y. Kato, N. Uchida, O. Fukuda, and K. Inada, Transmission characteristics of long spliced graded-index

optical fibers at 1.27 μm, *J. Quantum Electron.*, QE-15, 638 (1979).

12. D. Marcuse, *Theory of Dielectric Optical Waveguide*, Academic Press, New York, 1974, p. 60; K. Iga and Y. Kokubun, *Optical Fiber* [in Japanese], OHM-Sha, Tokyo, 1986, p. 58.

13. S. Seikai, M. Tateda, K. Kitayama, and N. Uchida, Optimization of multimode graded-index fiber parameters: design considerations, *Appl. Opt.*, 19, 2860 (1980).

14. M. Ohashi, K. Kitayama, Y. Koyamada, N. Uchida, and S. Seikai, Design of multimode graded-index fiber parameters in the long-wavelength region, *Appl. Opt.*, 23, 1802 (1984).

15. K. Kitayama, Y. Kato, M. Ohashi, Y. Ishida, and N. Uchida, Design considerations for the structural optimization of a single-mode fiber, *J. Lightwave Technol.*, LT-1, 363 (1983).

16. ITU-T. *Characteristics of a Single-Mode Optical Fibre Cable*, ITU-T Recommendation G.652, Geneva, Switzerland (October 2000).

17. L.G. Cohen, C. Lin, and W.G. French, Tailoring zero chromatic dispersion into the 1.5-1.6 μm low-loss spectral region of single-mode fibers, *Electron. Lett.*, 15, 334 (1979).

18. W.A. Gambling, H. Matsumura, and C.M. Ragdale, Zero total dispersion in graded-index single mode fibers, *Electron. Lett.*, 15, 474 (1979).

19. U.C. Paek, G.E. Peterson, and A. Carnevale, Dispersionless single-mode lightguides with index profile, *Bell Syst. Tech. J.*, 60, 583 (1981).

20. U.C. Paek, Dispersionless single-mode fibers with trapezoidal-index profiles in the wavelength region near 1.5 μm, *Appl. Opt.*, 22, 2363 (1983).

21. N. Kuwaki, M. Ohashi, C. Tanaka, and N. Uesugi, Dispersion-shifted convex-index single-mode fibers, *Electron. Lett.*, 21, 1186 (1985).

22. V.A. Bhagavatula, M.S. Spotz, and D.E. Quinn, Uniform waveguide dispersion segmented-core designs for dispersion-shifted single-mode fibers, *OFC*, MG2 (1984).

23. S. Kawakami and S. Nishida, Anomalous dispersion of new doubly-clad optical fiber, *Electron. Lett.*, 10, 38 (1974).

24. K. Okamoto, T. Edahiro, A. Kawana, and T. Miya, Dispersion minimizations in single-mode fibers over a wide spectral range, *Electron. Lett.*, 15, 729 (1979).

25. L.G. Cohen, W.L. Mammel, and S.J. Jang, Low-loss quadruple-clad single-mode lightguides with dispersion below 2psec/km·nm over the 1.28 µm-1.65 µm wavelength range, *Electron. Lett.*, 18, 1023 (1982).

26. ITU-T. *Characteristics of a Dispersion-Shifted Single-Mode Optical Fibre Cable*, ITU-T Recommendation G.653, Geneva, Switzerland (October 2000).

27. A.A. Griffith, The phenomena of rapture and flaw in solid, *Phys. Trans. R. Soc.*, 221, 163 (1920).

28. Y. Mitsunaga, Y. Katsuyama, H. Kobayashi, and Y. Ishida, Failure prediction for long length optical fiber based on proof testing, *J. Appl. Phys.*, 53, 4847 (1982).

29. D. Gloge, Bending loss in multimode fibers with graded and ungraded core index, *Appl. Opt.*, 11, 2506 (1972).

30. R. Olshansky, Distortion losses in cable optical fibers, *Appl. Opt.*, 14, 20 (1975).

31. N. Uchida, N. Uesugi, Y. Murakami, M. Nakahara, T. Tanifuji, and N. Inagaki, Infrared loss increase in silica optical fiber due to chemical reaction of hydrogen, *Ninth ECOC*, PD-1, 1 (1983).

32. N. Uesugi, Y. Murakami, C. Tanaka, Y. Ishida, Y. Mitsunaga, Y. Negishi, and N. Uchida, Infra-red optical loss increase for silica fiber in cable filled with water, *Electron. Lett.*, 19, 762 (1983).

33. J. Stone, A.R. Chraplyvy, and C.A. Burrus, Gas-in-glass—a new Raman-gain medium: molecular hydrogen in solid-silica optical fibers, *Opt. Lett.*, 7, 297 (1982).

34. N. Uesugi, T. Kuwabara, Y. Koyamada, Y. Ishida, and N. Uchida, Optical loss increase of phosphor-doped silica fiber at high temperature in the long wavelength region, *Appl. Phys. Lett.*, 43, 327 (1983).

35. N. Uchida and N. Uesugi, Infrared loss increase in silica fibers due to hydrogen, *J. Lightwave Technol.*, LT-4, 1132 (1986).

36. K. Hagimoto, K. Iwatsuki, A. Takada, M. Nakazawa, M. Satuwatari, K. Aida, K. Nakagawa, and M. Horiguchi, A 212 km non-repeated

transmission experiment at 1.8 Gb/sec using LD pumped Er^{3+}-doped fiber amplifiers in an IM/direct-detection repeater system, *12th OFC*, PD15 (1989).

37. K. Takada, M. Abe, T. Shibata, M. Ishii, Y. Inoue, H. Yamada, Y. Hibino, and K. Okamoto, 10 GHz-spaced 1010-channel AWG filter achieved by tandem connection of primary and secondary AWGs, *26th ECOC*, PD3.8 (2000).

38. R.W. Tkach, A.R. Chraplyvy, F. Forghieri, A.H. Gnauck, and R.M. Derosier, Four-photon mixing and high-speed WDM systems, *J. Lightwave Technol.*, 13, 841 (1995).

39. A. Hasegawa and F. Tappert, Transmission of stationary nonlinear optical pulses in dispersive dielectric fibers. I. Anomalous dispersion, *Appl. Phys. Lett.*, 23, 142 (1973).

40. K. Mori, H. Takara, S. Kawanishi, M. Saruwatari, and T. Morioka, Flatly broadened supercontinuum spectrum generated in a dispersion decreasing fibre with convex dispersion profile, *Electron. Lett.*, 33, 1806 (1997).

41. G.P. Agrawal, *Nonlinear Fiber Optics*, 2nd ed., Academic Press, San Diego, CA, 1995, p. 316.

42. Y. Emori, K. Tanaka, and S. Namiki, 100 nm bandwidth flat-gain Raman amplifiers pumped and gain-equalised by 12-wavelength-channel WDM laser diode unit, *Electron. Lett.*, 35, 1355 (1999).

43. J. Ishikawa and T. Chikama, Ultra high-speed TDM transmission technologies [in Japanese], *O plus E*, 21, 46 (1999).

44. F.P. Kapron, Systems considerations for polarization-mode dispersion, *Natl. Fiber Opt. Eng. Conf. (NFOEC'97)*, session 10, 433 (1997).

45. D.W. Peckham, A.F. Judy, and R.B. Kummer, Reduced dispersion slope, non-zero dispersion fiber, *24th ECOC*, TuA06, 139 (1998).

46. Y. Frignac and S. Bigo, Numerical optimization of residual dispersion in dispersion-managed systems at 40 Gb/sec, *OFC*, TuD3, 48 (2000).

47. Y. Liu, W.B. Mattingly, D.K. Smith, C.E. Lacy, J.A. Cline, and E.M. De Liso, Design and fabrication of locally dispersion-flattened large effective area fibers, *24th ECOC*, MoA05, 37 (1998).

48. ITU-T, *Characteristics of a Non-zero Dispersion Shifted Single-Mode Optical Fibre Cable.* ITU-T Recommendation G.655, Geneva, Switzerland (October 2000).

49. M. Hirano, T. Kato, T. Ishihara, M. Nakamura, Y. Yokoyama, M. Onishi, E. Sasaoka, Y. Makio, and M. Nishimura, Novel ring-core-dispersion-shifted fiber with depressed cladding and its four-wave mixing efficiency, *25th ECOC*, II.278 (1999).

50. K. Mukasa, Y. Akasaka, and Y. Suzuki, Development of segment core profile DSF with low non-linearity and dispersion slope, *Third Optoelectron. & Commun. Conf. (OECC)*, 15C1-4, 366 (1998).

51. P. Nouchi, P. Sansonetti, J. Von Wirth, and C. Le Sergent, New dispersion shifted fiber with effective area larger than 90 µm², *22nd ECOC*, MoB3.2, 1.49 (1996).

52. K. Aikawa, T. Suzuki, K. Himeno, and A. Wada, New dispersion-flattened hybrid optical fiber link composed of medium-dispersion large-effective-area fiber and negative dispersion fiber, *OFC*, TuH6 (2001).

53. Y. Liu and G. Berkey, Single-mode dispersion-shifted fibers with effective area over 100 µm², *24th ECOC*, MoA07, 41 (1998).

54. Y. Akasaka and Y. Suzuki, Enlargement of effective core area on dispersion flattened fiber and its low nonlinearity, *OFC*, ThK2, 302 (1998).

55. N. Kumano, K. Mukasa, S. Matsushita, and T. Yagi, Zero dispersion-slope NZ-DSF with ultra wide bandwidth over 300 nm, *28th ECOC*, PD1.4 (2002).

56. K. Mukasa, T. Yagi, and K. Kokura, Wide-band dispersion management transmission line with medial dispersion fiber (MDF), *26th ECOC*, 2.4.2, 93 (2000).

57. H. Izadpanah, C. Lin, J. Gimlett, H. Johnson, W. Way, and P. Kaiser, Dispersion compensation for upgrading interoffice networks built with 1310 nm optimized SMFs using an equalizer fiber, EDFAs and 1310/1550 nm WDM, *OFC*, PD15, 371 (1992).

58. M. Onishi, Y. Koyano, M. Shigematsu, H. Kanamori, and M. Nishimura, Dispersion compensating fiber with a high figure of merit of 250 psec/nm/dB, *Electron. Lett.*, 30, 161 (1994).

59. A.M. Vengsarkar, A.E. Miller, and W.A. Reed, Highly efficient single-mode fiber for broadband dispersion compensation, *OFC/IOOC*, PD13, 56 (1993).

60. M. Onishi, H. Kanamori, T. Kato, and M. Nishimura, Optimization of dispersion-compensating fibers considering self-phase modulation suppression, *OFC*, ThA2, 200 (1996).

61. Y. Akasaka, R. Sugizaki, A. Umeda, and T. Kamiya, High-dispersion-compensation ability and low nonlinearity of W-shaped DCF, *OFC*, ThA3, 201 (1996).

62. T. Tsuda, Y. Akasaka, S. Sentsui, K. Aiso, Y. Suzuki, and T. Kamiya, Broad band dispersion slope compensation of dispersion shifted fiber using negative slope fiber, *24th ECOC*, 233 (1998).

63. L. Grüner-Nielsen, T. Veng, S.N. Knudsen, C.C. Larsen, and B. Edvold, New dispersion compensating fibers for simultaneous compensation of dispersion and dispersion slope of non-zero dispersion shifted fibers in the C or L band, *OFC*, TuG6-1, 101 (2000).

64. M. Wandel, P. Kristensen, T. Veng, Y. Qian, Q. Le, and L. Grüner-Nielsen, Dispersion compensating fibers for non-zero dispersion fibers, *OFC*, WU1, 327 (2002).

65. L. Grüner-Nielsen, S.N. Knudsen, T. Veng, B. Edvold, and C.C. Larsen, Design and manufacture of dispersion compensating fiber for simultaneous compensation of dispersion and dispersion slope, *OFC*, WM13, 232 (1999).

66. T. Suzuki, K. Aikawa, K. Himeno, A. Wada, and R. Yamauchi, Large-effective-area dispersion compensating fibers for dispersion accommodation both in the C and L bands, *Fifth Optoelectron. & Commun. Conf. (OECC)*, 14C4-4, 554 (2000).

67. L. Grüner-Nielsen, S.N. Knudsen, B. Edvold, P. Kristensen, T. Veng, and D. Magnussen, Dispersion compensating fibres and perspectives for future developments, *26th ECOC*, 2.4.1, 91 (2000).

68. K. Mukasa, Y. Akasaka, Y. Suzuki, and T. Kamiya, Novel network fiber to manage dispersion at 1.55 µm with combination of 1.3 µm zero dispersion single mode fiber, *23rd ECOC*, MO3C, 127 (1997).

69. Y. Miyamoto, K. Yonenaga, S. Kuwahara, M. Tomizawa, A. Hirano, H. Toba, K. Murata, Y. Tada, Y. Umeda, and H. Miyazawa, 1.2-Tb/sec (30 × 42.7-Gb/sec ETDM optical channel) WDM transmission over 376 km with 125-km spacing using forward error correction and carrier-suppressed RZ format, *OFC*, PD26 (2000).

70. K. Mukasa and T. Yagi, Dispersion flat and low non-linear optical link with new type of reverse dispersion fiber (RDF-60), *OFC*, TuH7 (2001).

71. L.-A. de Montmorillon, F. Beaumont, M. Gorlier, P. Nouchi, L. Fleury, P. Sillard, V. Salles, T. Sauzeau, C. Labatut, J.-P. Meresse, B. Dany, and O. Leclerc, Optimized TeraLight™/reverse Tera-Light© dispersion- managed link for 40 Gb/sec dense WDM ultra long-haul transmission systems, *27th ECOC*, WeP44, 464 (2001).

72. K. Mukasa, K. Imamura, and T. Yagi, New type of positive medial dispersion fiber (P-MDF[150]) with dispersion as 10 psec/nm/km and A_{eff} about 150 μm^2, *OFC*, TuB1, 149 (2003).

73. K. Yonenaga and S. Kuwano, Dispersion-tolerant optical transmission system using duobinary transmitter and binary receiver, *J. Lightwave Technol.*, 15, 1530 (1997).

74. S. Bigo, Y. Frignac, G. Charlet, W. Idler, S. Borne, H. Gross, R. Dischler, W. Poehlmann, P. Tran, C. Simonneau, D. Bayart, G. Veith, A. Jourdan, and J.-P. Hamaide, 10.2 Tb/sec (256 × 42.7 Gb/sec PDM/WDM) transmission over 100 km TeraLight™ fiber with 1.28 b/sec/Hz spectral efficiency, *OFC*, PD25 (2001).

75. H. Sotobayashi, W. Chujo, and K. Kitayama, 1.6 b/sec/Hz, 6.4 Tb/sec OCDM/WDM (4 CDM × 40 WDM × 40 Gb/sec) transmission experiment, *27th ECOC*, PDM1.3, 6 (2001).

76. N.S. Bergano, C.R. Davidson, D.L. Wilson, F.W. Kerfoot, M.D. Tremblay, M.D. Levonas, J.P. Morreale, J.D. Evankow, P.C. Corbett, M.A. Mills, G.A. Ferguson, A.M. Vengsarker, J.R. Pedrazzani, J.A. Nagel, J.L. Zyskind, and J.W. Sulhoff, 100 Gb/sec error free transmission over 9100 km using twenty 5 Gb/sec WDM data channels, *OFC*, PD23 (1996).

77. K. Fukuchi, T. Kasamatsu, M. Morie, R. Ohhira, T. Ito, K. Sekiya, D. Ogasahara, and T. Ono, 10.92-Tb/sec (273 × 40-Gb/sec) triple-band/ultra-dense WDM optical-repeatered transmission experiment, *OFC*, PD24 (2001).

78. ITU-T, *Characteristics of a Cutoff Shifted Single-Mode Optical Fibre Cable*. ITU-T Recommendation G.654, Geneva, Switzerland (October 2000).

79. K.H. Chang, D. Kalish, and M.L. Pearsall, New hydrogen aging loss mechanism in the 1400 nm window, *OFC/IOOC*, PD22 (1999).

80. A. Dhillon, C. DiMinico, and A. Woodfin, *Optical Fiber and 10 Gb Ethernet*, White Paper, 10 Gb Ethernet Alliance (May 2002). Available at http//www.10gea.org/.

81. J.C. Knight, T.A. Birks, P.St.J. Russel, and D.M. Atkin, All-silica single-mode optical fiber with photonic crystal cladding, *Opt. Lett.*, 21, 1547 (1996).

82. K. Tajima, J. Zhou, K. Nakajima, and K. Sato, Ultra low loss and long length photonic crystal fiber, *OFC*, PD1 (2003).

8

Plastic Optical Fibers

TAKAAKI ISHIGURE

CONTENTS

8.1 INTRODUCTION

During the development of lightwave optical communications
with fiber, since the mid-1970s, the major emphasis for
research has been on the technology of lightwave devices for
long-distance telecommunications fields. Single-mode glass
optical fiber (GOF) is one of the most predictable and stable
communication channels ever developed and characterized for
such areas. The data communications market has risen to the
forefront even in lightwave communication because of the ever-
increasing need for more bandwidth. Ethernet is now used on
more than 80% of the world's personal computers (PCs) and
workstations connected to local-area networks (LANs) and
capability for priority-based transmission at 1000 Mb/sec
ensures that the Ethernet remains well ahead of other LAN
technologies. Much higher speed transmission (10 Gb/sec) has
been its challenging problem. As a physical-media-dependent
issue of not only Gigabit Ethernet but also 10-Gigabit Ethernet,
silica-based multimode fiber (MMF) is adopted to provide an
inexpensive optical link with a combination of transceivers based
on vertical cavity surface-emitting lasers (VCSELs). Therefore, it would not be necessarily the best solution to distrib-
ute such a silica-based optical fiber even in premises and
home networks or interconnections.

On the other hand, a plastic optical fiber (POF), with a
much larger core than that of silica fiber, has been expected
to be the office and home network media because its large core
allows the use of inexpensive injection-molded plastic connectors, which can dramatically decrease the total link cost. The
high-bandwidth graded index (GI) POF was proposed for the
first time[1] and its bandwidth characteristics have been
reported.[2,3] Polymethylmethacrylate (PMMA) has generally
been used as the core material of commercially available step-
index-type (SI) POF, and its attenuation limit is approximately 100 dB/km at the visible region.[2] Therefore, high

attenuation of the POF compared to the silica-based fiber has been one of the big barriers for POF in data communications applications of more than 100 m.

The development of the perfluorinated (PF) amorphous polymer based GI POF[4] opened the way for advantages in the high-speed POF network. Because serious intrinsic absorption loss caused by the carbon hydrogen vibration that existed in PMMA-based POF was completely eliminated in the PF-polymer-based POF, the experimental total attenuation of the PF-polymer-based GI POF decreased to 10 dB/km even in the near-infrared region.[4] It was clarified that the theoretical attenuation limit of the PF-polymer-based POF is more comparable to that of the silica-based fiber (0.3 dB/km).

It was theoretically found that by forming an optimum refractive index profile in the PF-polymer-based GI POF, more than 10-Gb data transmission is possible over 1 km because the PF-polymer-based GI POF has such advantages as low intrinsic loss and low material dispersion.[5] For silica-based MMF fabricated by the conventional modified chemical vapor deposition (MCVD) process, it is generally known that modal dispersion dominates its bandwidth characteristic because of the refractive index profile perturbation, such as a central dip.[6,7] It is well known that the ideal refractive index profile in the MMF provides the minimum modal dispersion.[3] Experimentally obtained GI POF with the ideal index profile showed much higher bandwidth that was almost independent of the launch condition.

8.2 DEVELOPMENT OF PLASTIC OPTICAL FIBERS FOR HIGH-SPEED DATA COMMUNICATION

The first report of POF was in the 1960s, which was almost the same time as the invention of silica-based optical fiber. However, POF has not necessarily played a main role in the field of telecommunications because of its intrinsic high attenuation. PMMA, which has been the typical polymer material for the core of POF, shows the lowest attenuation at the visible region, and the attenuation abruptly increases

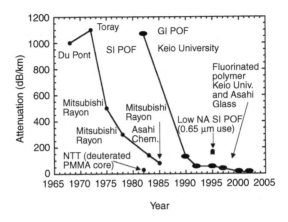

Figure 8.1   Development of POF attenuation. NTT, Nippon Telegraph and Telephone.

from near-infrared (IR) to IR regions. This is because the intrinsic absorption loss from carbon–hydrogen stretching vibration. Therefore, the theoretical attenuation limit of PMMA-based POF is around 100 dB/km at a 0.65-μm wavelength. However, in the 1990s PF polymer, an innovative polymer material, was invented.[8] As the PF polymer has no carbon–hydrogen bonding, the intrinsic absorption loss dramatically decreases.[9] Development of POF attenuation is summarized in Figure 8.1.

With decreasing attenuation of POF, high bandwidth also became a requirement for POF. All commercially available POFs have been of step index SI POF. Therefore, the bit rate of the conventional POF link has been limited to less than 10 Mb/sec; the 100-Mb/sec data rate was achieved in the SI POF link by decreasing its modal dispersion by a low numerical aperture (NA).[10] On the other hand, a high-bandwidth GI POF was proposed, and in 1994, it demonstrated a 2.5-Gb/sec, 100-m transmission.[11] Proposal of the GI POF has triggered the research and development of a high-speed POF data link. Experimental results for high-speed data transmission by POF are shown in Figure 8.2. Using low-loss PF-polymer-based GI POF, 2.5-Gb/sec, 450-m transmission succeeded in 1998.

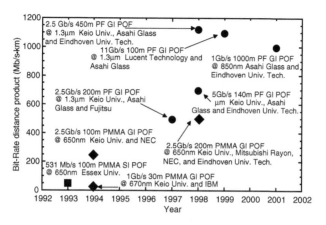

Figure 8.2 Development of high-speed POF link.

8.3 ADVANTAGE OF LARGE-CORE GRADED INDEX PLASTIC OPTICAL FIBERS

8.3.1 Coupling Loss

8.3.1.1 Light Source to Fibers

One of the most remarkable advantages of the POF is its large core diameter, which maintains great flexibility. Because of the large core, the light-coupling efficiency between the light source and fiber and fiber-to-fiber is dramatically improved, which decreases the cost of optical connectors for POFs. Regarding the coupling losses of POFs, the suitable core diameter of the GI POF for high coupling efficiency with a light source was estimated by Nyu et al.[12] In their estimation, the current low-cost POF transceivers in which a red laser diode (LD) or light-emitting diode (LED) was bonded onto a lead frame and molded in a clear plastic package was supposed to be used. When the LD was adopted as the light source, it was reported that the coupling loss mainly depended on the radiation angle of the LD; namely, the radiation angle of the light source was a more important factor than the core diameter of the GI POF. The coupling loss between the LD and the GI POF is listed

TABLE 8.1   Calculated Coupling Loss between an LD and a GI POF

Radiation angle of an LD	Core diameter = 600 μm	Core diameter = 1000 μm
27°	3.0 dB	1.3 dB
17°	1.3 dB	0.4 dB

in Table 8.1. Use of a 600-μm core diameter GI POF with the LD with a 17° radiation angle realizes approximately 1 dB of coupling loss.

8.3.1.2   Fiber-to-Fiber

Theoretical coupling loss of two GI-type optical fibers is

$$L(dB) = -10\log\left\{\cos^{-1}\left(\frac{d}{2a}\right) - \left[1 - \left(\frac{d}{2a}\right)^{1/2}\right]\left(\frac{6a}{d}\right)\left\{5 - \left(\frac{d^2}{2a^2}\right)\right\}\right\}$$

(8.1)

where d is the axial displacement, and a is the core radius. It is assumed that all rays are uniformly excited. Both fibers have the same dimensions and are of the same type.

Table 8.2 shows the estimated coupling losses between two GI glass fibers or polymer optical fibers by Equation 8.1. Here, the core diameters of the GOF and POF were defined

TABLE 8.2   Comparison of Calculated Coupling Loss in the Two Fiber Connection between the GI GOF and GI POF

Amount of displacement $d(\mu m)$	GI GOF Core diameter = 60 μm	GI POF Core diameter = 600 μm
2.5	0.37 dB	0.040 dB
5.0	0.71 dB	0.079 dB
7.5	1.04 dB	0.12 dB
50	5.44 dB	0.71 dB
75	—	1.04 dB

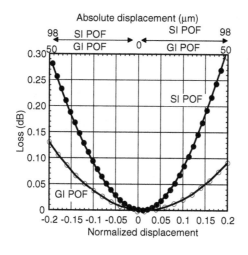

Figure 8.3 Coupling losses between two POFs.

as 50 and 500 μm, respectively. It should be noted that, for the GI POF, a slight amount of displacement (e.g., 5 to 10 μm) causes negligible coupling loss.

Figure 8.3 shows the experimental results of the connection loss of two GI POFs with a 500-μm core compared with that of the low-NA-type SI POF with a 980-μm core. The NA of the GI POF is 0.28, which is almost the same as that of the low-NA SI POF. It should be emphasized that the GI POF has a big advantage in the connection loss of fibers when a slight amount of displacement is caused at the connecting point. As shown in Figure 8.3, the connection loss was almost the same as that of the SI POF, although the GI POF had almost half the core diameter compared with the SI POF. This is because of the difference of near-field output power intensity distribution between two fibers. In the SI POF, the output power is uniform in the whole core region; the near-field power of the GI POF is maximum at the core center, and it gradually decreases to the periphery. Therefore, the tolerance of displacement between GI POFs increases, and the slight amount of displacement is serious for power coupling in the connection of SI POFs.

8.3.2   Modal Noise Elimination

It was shown in previous work on silica-based MMFs with
50- to 62.5-μm core diameters that modal noise deteriorated
system performance.[13] Modal noise is caused when a MMF
is utilized with coherent light sources such as LDs or VCSELs.
A speckle pattern is formed by interference between the
propagating modes on the output end of the fiber. If an offset
connection between two fibers is included in the MMF link,
the speckle power distribution is translated into intensity fluctuation, which is the modal noise. In the case of 50- or 62.5-μm
core silica-based MMF, the permissible misalignment in the
connector is several micrometers, although its larger core diameter compared to the single-mode counterpart tolerates the
misalignment in the fiber connection from the aspect of the
connection loss. On the other hand, such a small misalignment
is negligible from the aspect of the connection loss for the GI
POF with a much larger core (>120 μm). In addition to the
advantage in connection loss, it was clarified that the modal
noise was virtually eliminated in the large-core (>120 μm) GI
POF because of the huge number of propagating modes.

Figure 8.4 shows the effect of the modal noise on the eye
pattern of a 50-μm core GI MMF link. In this experiment, an
output light from an LD transmitter at a 0.65-μm wavelength

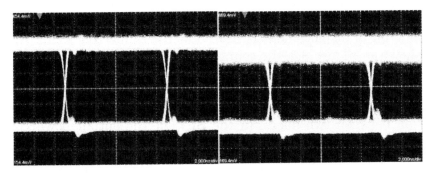

(a) No displacement   (b) 20-μm displacement

Figure 8.4   Modal noise effect on eye diagram of a 100-Mb/sec data
transmission by silica-based multimode fiber. (a) no displacement
at the fiber connection point; (b) a 20-μm displacement.

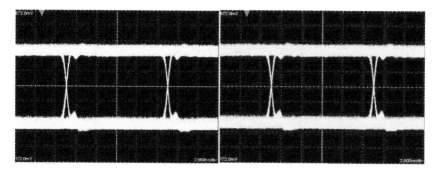

(a) No displacement   (b) 60-μm displacement

Figure 8.5 Modal noise effect on eye diagram of a 100-Mb/sec data transmission by graded index plastic optical fiber (GI POF) with 120-μm core: (a) no displacement at the fiber connection point; (b) a 60-μm displacement.

was coupled into a 2-m incoming fiber. Another 2-m outgoing fiber was connected to the incoming fiber on a V-groove that was placed on a triaxial manipulator. Output light from the outgoing fiber was coupled to a Si-photodiode. At this connection point, two fibers could be deliberately misaligned to cause a mode-selective loss (MSL). As shown in Figure 8.4, a slight amount of MSL causes serious deterioration of the eye diagram for silica-based MMF. Therefore, micrometer-order alignment is necessary in the fiber connection even if the core diameter is much larger than that of single-mode fiber.

In contrast, the modal noise effect in the GI POF with a 120-μm core is shown in Figure 8.5. It was confirmed that the larger core of the GI POF enables complete elimination of the modal noise problem on the eye diagram even with 60-μm displacement.

8.4 LOW-LOSS PLASTIC OPTICAL FIBERS

PMMA has been used for the core material of POF to date because it has been recognized as a highly transparent polymer with commodity. Great efforts to lower the attenuation of the POF were devoted in the 1980s by investigation of the

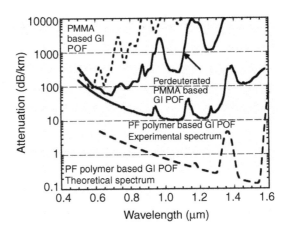

Figure 8.6   Total attenuation spectra of GI POFs.

type of polymer material and by improvement of the purification process of the materials.[14] The attenuation of transmission of the PMMA-based GI POF is shown in Figure 8.6. The minimum attenuation was about 150 dB/km at a 0.65-μm wavelength, which was almost the same as that of the step-index-type POF commercially available. However, the attenuation of PMMA-based POF abruptly increased from about a 0.6-μm wavelength to the IR region because of the absorption loss of overtones of carbon–hydrogen stretching vibration. On the other hand, it is highly desirable to construct a POF network system using commercially available LDs and LEDs that operate in the 0.6- to 1.5-μm wavelength range. Deuterated or fluorinated polymer-based POF will be a promising candidate to eliminate the serious absorption loss in such wavelengths.

The attenuation spectra of perdeuterated and PF polymer-based GI POFs are also shown in Figure 8.6. As basic chemical properties such as polymerization reactivity of the deuterated monomer are almost the same as that for the hydrogenated counterpart, the same fabrication process as for the PMMA-based GI POF can be applied to the deuterated methylmethacrylate (MMA) monomer. As the refractive index of the PMMA-d8 is almost the same as that of PMMA, the cladding, where little optical power is transmitted, could consist of PMMA. Therefore, the PMMA-d8 material is used only for the core region, which is cost-effective compared with the PF polymer. The attenuation spectrum of the PMMA-d8-based GI-POF at a 0.65-μm wavelength (58 dB/km) is lower than that (152 dB/km) of the PMMA-based GI-POF. In addition to the low attenuation at a 0.65-μm wavelength, the wide low-loss optical window from a 0.63- to 0.68-μm wavelength is the large advantage of the PMMA-based GI POF. Therefore, considering the wavelength shift of a light source such as an LD or VCSEL by the operating temperature change, the low-loss wide optical window near the 0.65-μm wavelength in the spectrum deuterated PMMA-based GI POF is advantageous.

On the other hand, it is noteworthy that the PF-polymer-based GI POF has no serious absorption peak in the 0.5- to 1.3-μm wavelength range, and the attenuation even at the near-IR region is about 10 dB/km. Therefore, a light source not only at the visible region but also at the near-IR region can be used for a PF GI POF link.

Total attenuation spectrum of the PF-polymer-based POF was estimated by summation of scattering and absorption losses. The result is shown in Figure 8.6. Here, it was assumed that the absorption peak at 1.361 μm was the Gaussian profile with 0.020-μm full-width half-maximum. It is indicated that the attenuation limit of the PF-polymer-based GI POF at a 1.3-μm wavelength is approximately 0.3 dB/km, which is comparable with that of silica fiber (0.2 dB/km).

8.5 HIGH-BANDWIDTH PLASTIC OPTICAL FIBERS

In the early 1990s, there were a few reports regarding the high-bit-rate transmission by POF because there were no appropriate devices (e.g., LD and photodetector) for POF. Because a 2.5-Gb/sec data transmission was reported by Keio University and NEC in 1994, much interest has been focused on the POF data link.[15,16] With the development of the low-loss PF-polymer-based GI POF, the bit rate–distance product increased (Figure 8.4). In 1999, an outstanding demonstration was reported by Bell Laboratory, Lucent Technologies[16] regarding experimental 11-Gb/sec data transmission by 100-m PF-polymer-based GI POF at a 1.3-μm wavelength. In this section, the bandwidth property of the GI POF is described.

It is well known that the modal dispersion of the multimode optical fiber can be minimized by optimizing the refractive index profile of the core region. Optimization of the dispersions of GI POF not only for modal dispersion but also for material and profile dispersions is described. To design the optimum index profile of the GI POF, the index profile is approximated by the power law form described in Equation 8.2:

$$n(r) = n_1 \left[1 - 2\Delta\left(\frac{r}{a}\right)^g \right]^{1/2} \tag{8.2}$$

$$\Delta = \frac{n_1^2 - n_2^2}{2n_1^2} \approx \frac{n_1 - n_2}{n_1} \tag{8.3}$$

Here, n_1 and n_2 are refractive indices at the center axis and cladding of the fiber, respectively; a is the radius of the core; and Δ is the relative difference of the refractive index. The profile of the refractive index was evaluated by the parameter g, called the index exponent. The material and profile dispersions of the POF were estimated by measuring the wavelength dependence of the refractive index of polymers.[5,17] The results of the material dispersion are shown in Figure 8.7, which was derived from the data of wavelength dependence of the refractive index using Equation 8.4:

$$D_{mat} = -\left(\frac{\lambda\delta\lambda}{c}\right)\left(\frac{d^2n}{d\lambda^2}\right)L \tag{8.4}$$

where $\delta\lambda$ is the root mean square spectral width of the light source, λ is the wavelength of transmitted light, c is the velocity of light in a vacuum, $d^2n/d\lambda^2$ is the second-order dispersion, and L is the length of the fiber. It is noteworthy that the material dispersion of the PF polymer is smaller than that of silica in the near-IR region.

Figure 8.8 shows the relation between the bandwidth characteristics and refractive index profile of the PMMA- and

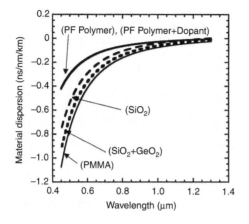

Figure 8.7 Material dispersions of PMMA, silica, and PF polymer.

PF-polymer-based GI POF calculated by the Wentzel-Kramer-Brillouin (WKB) method,[17,18] in which both modal and material dispersions were taken into consideration. Here, the light source was assumed to be an LD with a 3-nm spectral width. In the case of PMMA-based GI POF, the optimum wavelength at which the lowest attenuation can be obtained is located at

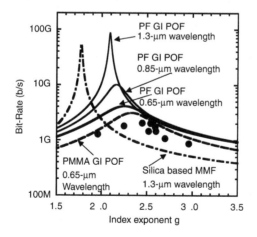

Figure 8.8 Calculated relation between the index exponent of the GI POF and possible bit-rate in the GI POF link. Closed circle, experimental value.

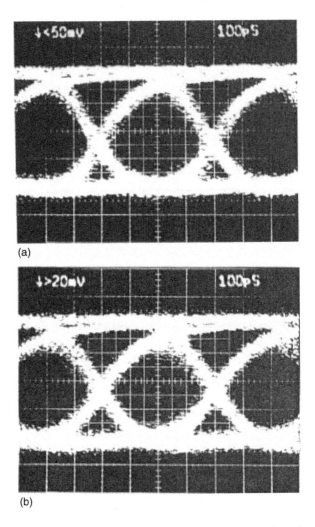

(a)

(b)

Figure 8.9   Received waveform at 647-nm wavelength and 2.5-Gb/sec transmission.

0.65 μm. Eye diagrams after a 2.5-Gb/sec 100-m PMMA-based GI POF transmission are shown in Figure 8.9. A good eye opening is observed.[11] The maximum bandwidth of the PMMA-based GI POF is around 3 Gb/sec for 100 m, which is dominated by a large material dispersion shown in Figure 8.7.

On the other hand, the low material dispersion of PF polymer enables 9-Gb/sec transmission in a 100-m link at the same wavelength. Furthermore, the low intrinsic absorption loss of the PF-polymer-based GI POF permits 1.3-μm wavelength use. Because the material dispersion decreases with an increase in wavelength, the possible bit rate in the 100-m link achieves 100 Gb/sec. It is noteworthy that the value of 100 Gb/sec is higher than the maximum bit rate of the silica fiber at the 1.3-μm wavelength because the material dispersion of the PF polymer is smaller than that of the silica material.[5]

Calculated bandwidth properties of the PF-polymer-based GI POF with respect to the wavelength are shown in Figure 8.10. For PF-polymer-based GI POF, several types of dopants were investigated to represent the high performances, such as low attenuation, high mechanical strength, high-temperature resistance, and so on. In the bit rate calculation process, the material dispersion of the PF polymer including the best dopant to satisfy the requirements of the POF was taken into consideration. In Figure 8.10, it was assumed that the refractive index profiles of both silica-based MMF and PF-polymer-based GI POF were designed to be optimum for 0.85 μm use. It was confirmed that if the index exponent g is controlled to around 2.0, several gigabits per second for

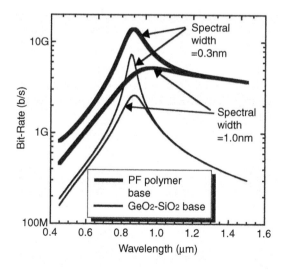

Figure 8.10 Wavelength dependence of the possible bit-rate in PF GI POF link compared with that in silica-based multimode fiber link.

100 m could be achieved by PF-polymer-based GI POF in the wide-wavelength range from 0.6 to 1.3 μm. On the other hand, in the case of the SiO_2–GeO_2-based MMF, accurate index profile control for a specified wavelength is necessary to achieve several gigabits per second for 100 to 300 m because the wavelength dependence of the bandwidth is much larger than that of the PF polymer, as shown in Figure 8.10.

In Figure 8.10, the bit rate performance of the PF-polymer-based GI POF link was also shown when the spectral source width was assumed to be 0.3 nm, which corresponded to the VCSEL, while 1 nm of spectral width corresponded to a Fabry–Perot-type laser. It is obvious that with decreasing spectral width of the light source, the maximum bandwidth at the 0.85-μm wavelength increases, and more than 10 GHz•km of bandwidth can be achieved. This is because the effect of material dispersion is more serious in the short-wavelength range.

As shown in Figure 8.10, in the case of PF-polymer-based GI POF, the use of VCSELs with a narrow line width enables transmission at more than a gigabit in a wide wavelength range (0.6 to 1.6 μm), which cannot be realized by the silica-based MMF with a larger material dispersion than PF polymer. Furthermore, because the PF-polymer-based GI POF has low attenuation in the wide-wavelength range from the visible to near-IR region, not only a serial high-speed data transmission system but also a WDM system can be realized. When the coarse WDM system is applied, the wavelength dependence of the bit rate performance observed in the silica-based MMF link was a concern in the POF link. However, if the PF-polymer-based GI POF is fabricated to have an approximately optimum index profile, it can be expected that results higher than the order of gigabits per second for 1 km can be covered in the 0.8- to 1.3-μm wavelength region. Such a high bit rate cannot be achieved by the silica-based MMF link in the wide-wavelength range.

8.6 SUMMARY AND FUTURE PROSPECTS

Progress of the GI POF, particularly its bandwidth, attenuation, and reliability, was reviewed. It is generally believed that high optical and mechanical quality cannot be realized by polymer but by inorganic materials in the photonics fields. However, it was clarified for the first time that the PF-polymer-based GI POF potentially had much higher bandwidth than silica because the material dispersion of the PF polymer was lower than that of silica. The attenuation of the PF-polymer-based GI POF can be decreased to less than 1 dB/km, which is comparable with that of the silica-based fiber. The low attenuation and low material dispersion of the PF-polymer-based GI POF present a solution to the ever-increasing bandwidth demand in the interconnections and access networks.

REFERENCES

1. T. Ishigure, E. Nihei, and Y. Koike, Graded-index polymer optical fiber for high-speed data communication, *Appl. Opt.*, 33, 4261–4266 (1994).

2. Y. Koike, T. Ishigure, and E. Nihei, High-bandwidth graded-index polymer optical fiber, *IEEE J. Lightwave Technol.*, 13, 1475–1489 (1995).

3. T. Ishigure, S. Tanaka, E. Kobayashi, and Y. Koike, Accurate refractive index profiling in a graded-index plastic optical fiber exceeding gigabit transmission rates, *IEEE J. Lightwave Technol.*, 20, 1449–1456 (2002).

4. Y. Koike and T. Ishigure, Bandwidth and transmission distance achieved by POF, *IEICE Trans. Electron.*, E82-C, 1287–1295 (1999).

5. T. Ishigure, Y. Koike, and J.W. Fleming, Optimum index profile of the perfluorinated polymer based GI polymer optical fiber and its dispersion properties, *IEEE J. Lightwave Technol.*, 18, 178–184 (2000).

6. M. Webster, L. Raddatz, I.H. White, and D.G. Cunningham, A statistical analysis of conditioned launch for Gigabit Ethernet links using multimode fiber, *IEEE J. Lightwave Technol.*, 17, 1532–1541 (1999).

7. L. Raddatz, I.H. White, D.G. Cunningham, and M.C. Nowell, An experimental and theoretical study of the offset launch technique for the enhancement of the bandwidth of multimode fiber links, *IEEE J. Lightwave Technol.*, 16, 324–331 (1998).

8. K. Oharu, N. Sugiyama, M. Nakamura, and I. Kaneko, Preparation and reaction of perfluoro(alkenyl vinyl ether), *Rep. Res. Lab. Asahi Glass Co. Ltd.*, 41, 51 (1991).

9. W. Groh, D. Lupo, and H. Sixl, Polymer optical fibers and nonlinear optical device principles, Angew. *Chem. Int. Ed. Engl. Adv. Mater.*, 28, 1548–1559 (1989).

10. S. Yamazaki, High speed plastic optical fiber transmission for desk-top LAN, *21st European Conf. Opt. Commun. (ECOC'95)*, 1, 337–343 (1995).

11. T. Ishigure, E. Nihei, K. Kobayashi, S. Yamazaki, and Y. Koike, 2.5 Gb/sec 100 m data transmission using graded-index polymer optical fiber and high-speed laser diode at 650nm wavelength, *Electron. Lett.*, 31, 467–468 (1994).

12. T. Nyu, S. Yamazaki, and Y. Koike, Design for Core Diameter of Wide-band GI-POF, Proceedings of Plastic Optical Fibers and Applications Conference, POF'95, Boston, October 1995.

13. R.E. Epworth, The phenomenon of modal noise in analogue and digital optical fiber systems, *Proc. Fourth Eur. Conf. Opt. Commun.*, Genova, Italy, 492–501 (September 1978).

14. T. Kaino, M. Fujiki, S. Oikawa, and S. Nara, Low-loss plastic optical fibers, *Appl. Opt.*, 20, 2886–2888 (1981).

15. W. Li, G.D. Khoe, H.P.A. v.d. Boom, G. Yabre, H. de Waardt, Y. Koike, M. Naritomi, N. Yoshihara, and S. Yamazaki, Record 2.5 Gb/sec Transmission Via Polymer Optical Fiber at 645 nm Visible Light, Seventh International POF Conference, Post Deadline Paper, Berlin, October 1998.

16. G. Giaretta, W. White, M. Wegmueller, R.V. Yelamatry, and T. Onishi, 11 Gb/sec Data Transmission Through 100 m of Perfluorinated Graded-Index Polymer Optical Fiber, Optical Fiber Communication Conference, Post Deadline Paper, PD14, San Diego, CA, February 1999.

17. J.W. Fleming, Material and mode dispersion in $GeO_2 \cdot B_2O_3 \cdot SiO_2$ glasses, *J. Am. Cer. Soc.*, 59, 503–507 (1976).

18. R. Olshansky and D.B. Keck, Pulse broadening in graded-index optical fibers, *Appl. Opt.*, 15, 483–491 (1976).

9

Optical Fiber Jisso Technology

NOBUO HORI

CONTENTS

9.1 INTRODUCTION

The Jisso technology of optical fibers involves various techniques, such as alignment, fixing, packaging, and various inspection techniques. Coupling optics that need the optical fiber Jisso technology include various types of systems, such as fiber/optical waveguide, laser diode (LD)–fiber, fiber–photodiode (PD), and fiber–fiber coupling. It is necessary to optimize the technique according to the bit rate, coupling efficiency, long-term reliability, and cost required in each coupling optics. The structures of coupling optics to which the optical fiber Jisso technology is applied are summarized here, as are alignment, fixing, and inspection used for those structures. In addition, packaging methods are briefly described.

9.2 STRUCTURES OF COUPLING OPTICS USING OPTICAL FIBERS

9.2.1 Direct Coupling

Direct coupling is a method of directly joining an optical fiber to a light-emitting or light-receiving device after bringing the fiber close to the device. Because the maximum coupling efficiency obtained by bringing a single-mode fiber and a light-emitting device as close as possible to each other is less than 10%, coupling a light-emitting device and an optical fiber by this method is limited mainly to situations that use MMFs.[1]

When direct coupling between a light-receiving device and an optical fiber is considered, if the device has a considerably large light-sensitive area, this method can be applied. However, if emphasis is placed on the efficiency of using a device that shows a high-speed response and has a small light-sensitive area, it is necessary to use a coupling lens.

It can be said that this method is the most economical coupling means for the following reasons: The coupling efficiency is intrinsically low; high fiber alignment or fixing accuracy is not necessary; there are fewer elements.

9.2.2 Lens Coupling

In the case of optical coupling from an LD to a single-mode fiber, the mode field diameter (MFD) of the LD differs from that of the fiber by a factor of 3 to 6. Therefore, it is necessary to use optics that gives an optimum imaging magnification.

$$\beta = \frac{\omega}{\sqrt{\omega_x^2 \cdot \omega_y^2}} \tag{9.1}$$

$$\eta = \frac{4\beta^2 \omega^2 \omega_x \omega_y}{\left(\beta^2 \omega_x^2 + \omega^2\right)\left(\beta^2 \omega_y^2 + \omega^2\right)} \tag{9.2}$$

The optics used here is applied when a single lens is employed and when a combination of two collimator lenses (LD collimator lens and fiber collimator lens) is employed (Figure 9.1). When a single lens is used and sufficient coupling efficiency is obtained, the lens needs to be selected so it is aberration corrected to minimize the aberration with an optical arrangement that gives the desired imaging magnification. When a pair of collimator lenses is used, the imaging magnification is optimized by the combination of the focal length of the lenses. When an isolator is used to suppress wavelength variations and variations in the output of an LD caused by reflected and returning light, the isolator is inserted between the lens

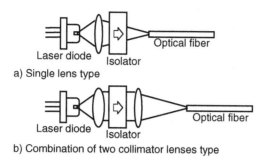

a) Single lens type

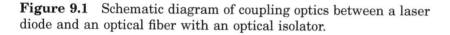

b) Combination of two collimator lenses type

Figure 9.1   Schematic diagram of coupling optics between a laser diode and an optical fiber with an optical isolator.

and the fiber in the case of a single lens and between the two collimator lenses in the case of a pair of collimator lenses.

In the case of coupling between single-mode fibers, the optics is constructed using identical collimator lenses such that the optics is an imaging system with unity magnification. A functional element such as a filter, isolator, circulator, or attenuator is positioned in the collimator portion between the lenses (Figure 9.2).

The interface of each optical element that is in contact with air, including the end faces of fibers, is antireflection (AR) coated. When the antireflection is improved, the fiber end face is polished obliquely, and then it is AR coated. This structure can suppress the return loss more than 60 dB.

9.2.3   Butt Joint

When an optical fiber and an optical waveguide are coupled, their end faces are brought into direct contact. This is referred to as a *butt joint*. In the present method, if the fiber and optical

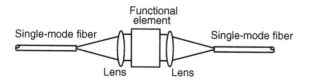

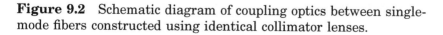

Figure 9.2   Schematic diagram of coupling optics between single-mode fibers constructed using identical collimator lenses.

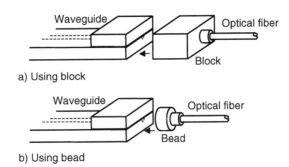

a) Using block

b) Using bead

Figure 9.3 Schematic view of butt jointing between a waveguide and an optical fiber.

waveguide are almost equal in MFD, high coupling efficiency can be obtained. In other cases, however, it cannot be corrected, and a connection loss corresponding to the difference in MFD occurs. In the case of a quartz-based optical waveguide, the coupling loss with the fiber can be suppressed to below –0.1 dB. In the case of a quartz-based optical waveguide and a quartz-based optical fiber, the return loss can be suppressed to better than 40 dB by coupling them using an adhesive matched with quartz in terms of refractive index. Normally, the return loss can be further suppressed to better than 60 dB by obliquely polishing the end faces of the fiber and waveguide at the same angle. In this method, the adhesive is interposed in the optical path. Therefore, when high-output light is used, it is necessary to pay attention to the durability of the adhesive. In a structure in which coupling to a waveguide substrate is made only with an optical fiber, the adhesive area is small, so sufficient adhesive strength cannot be obtained. Normally, contrivances are made to enhance the coupling strength. That is, the fiber is held in a block or bead provided with a fiber guide hole (Figure 9.3).[2] The end face is polished.

9.2.4 Various Types of Machining of Optical Fiber Tips

9.2.4.1 Polishing

When an optical fiber is used, it is necessary that its end face be optically polished such that light can go in and out of the

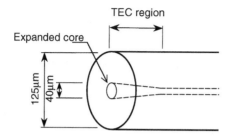

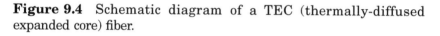

Figure 9.4 Schematic diagram of a TEC (thermally-diffused expanded core) fiber.

end face efficiently. When the effects of reflected and returning light present problems, the polishing is done at an angle to prevent recoupling of the light reflected at the end face into the fiber core.

9.2.4.2 Antireflective Coating

With respect to optical fibers used in modules, their end faces are AR coated to suppress reflected light and to prevent deterioration of the transmission efficiency.

9.2.4.3 Thermally-Diffused Expanded Core Fiber

For a thermally-diffused expanded core (TEC) fiber, the end of a single-mode fiber is heated to diffuse the core dopant into the cladding such that the MFD increases gradually toward the end (Figure 9.4). Single-mode light expanded to about 30 μm can be taken out with almost no loss. This can greatly improve the axial misalignment tolerance of the coupling efficiency of a coupling optical system.[3] In addition, this can also be used as a collimator with a short working distance.

9.2.4.4 Lensed Fiber

A spherical lensed fiber fabricated by machining the tip of an optical fiber with a 5- to 20-μm radius of curvature is used for direct coupling to an LD. Consequently, the coupling efficiency is improved compared with direct coupling. In a reported example, a coupling loss of –2 dB has been achieved using a lensed fiber with R = 10 μm for direct coupling to an LD. Furthermore, the number of elements is reduced compared with the case for which lenses are used. Thus, cost savings can be accomplished. A cylindrical lensed fiber obtained by machining only one side of an optical fiber into a wedge-shaped form, or lenticular form, is effective in enhancing the coupling efficiency by correcting the astigmatic difference of an LD.

It is also possible to achieve a fiber-to-fiber connection loss of 0.5 dB with a fiber collimator. Using a graded index multimode fiber (GI MMF) as a gradient index (GRIN) lens, the MMF is fusion spliced to the single-mode fiber end. Generally, lensed fibers have small collimated beam diameters of about 50 μm, so their working distances are limited to about 100 μm. However, the number of components can be reduced. This leads to cost reduction. Furthermore, high stability is obtained because they are fabricated by fusion without using dissimilar materials or by direct machining for optical fiber as shown in Figure 9.5.

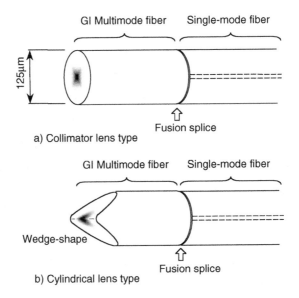

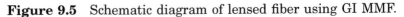

Figure 9.5   Schematic diagram of lensed fiber using GI MMF.

9.3 ALIGNMENT TECHNIQUES FOR OPTICAL FIBERS IN COUPLING OPTICS

9.3.1 Active Type

The technique for aligning the optical fiber to maximize optical coupling efficiency while monitoring it as the amount of light coupled to the input fiber is known as *active alignment.*

Companies have variously contrived algorithms for alignment, ranging from rough alignment to fine alignment, to improve alignment speed. A method generally adopted consists of rough and fine searches. That is, in the rough search a coordinate at which the amount of coupled light maximizes is found roughly. In the fine search, a peak coordinate is strictly identified on the order of submicrons. More specifically, in the rough search a scan is made in a zigzag or spiral manner. In the fine search, a cross scan is repeated around the coordinate showing a maximum amount of coupled light to seek for a new coordinate showing a maximum amount of coupled light (Figure 9.6).

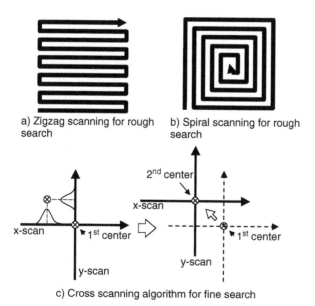

a) Zigzag scanning for rough search

b) Spiral scanning for rough search

c) Cross scanning algorithm for fine search

Figure 9.6 Scanning algorithm in active alignment for optimization of optical coupling.

The fiber alignment ends when the distance between the newly found coordinate showing a maximum amount of coupled light and the previous coordinate showing a maximum amount of coupled light converges to less than a given value. Genetic algorithms have been adopted in some cases to improve the alignment speed and to avoid convergence at suboptimal solutions. However, the most effective means for improving the alignment speed consists of increasing the dimensional accuracy of each member and the accuracy of the positions of elements held to the members and narrowing the searched range as much as possible during fiber alignment.

It is necessary to make the contact surfaces of the members parallel before fiber alignment to avoid deviation of the optical axis because of movement of a member when it is brought into contact after fiber alignment or it is fixed in the next bonding step. In one active method of achieving the parallel geometry, a holder for holding one member sits on an air swivel. The members are brought into contact. When their contact surfaces are aligned, the air swivel is locked. In another active method, when two members are in contact, one member is tilted. The amount of movement of the member in the abutting direction corresponding to the angle of tilt is monitored. The swivel is locked at an extreme value of the amount of movement. In another method adopted, the gap between members is observed from two mutually perpendicular directions. Image processing is performed to achieve parallel geometry. In some cases, however, sufficient accuracy is not obtained because of slight deviation of the accuracy at which the member edges are machined or because of slight deviation of the directions of observation.

9.3.1.1 LD–Fiber Coupling

A method of aligning the fiber in the case of single lens coupling is described here. In coupling optics between an LD and an optical fiber, a coupling efficiency is secured using an imaging optical system with a magnification ×3 to 5 because of the difference in MFD. In this case, misalignment between the LD and the lens produces a spot position deviation of ×3 to 5

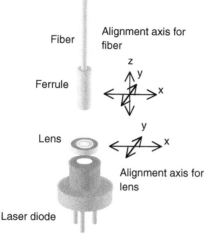

Figure 9.7 Schematic view of alignment assignment for cylindrical-type laser.

on the fiber end face with respect to the axial misalignment between the lens and fiber. Therefore, mechanical fixing between the LD and the lens is done first. The deviation of the spot position produced during this mechanical fixing is corrected by realigning the fiber. Finally, the position of the fiber is fixed as shown in Figure 9.7. Limitations are placed on the maximum coupling efficiency between the LD and the fiber because of asymmetry of the divergence angle of the output light from the LD. As the asymmetry of the divergence angle increases, the coupling efficiency decreases.

9.3.1.2 Fiber–Waveguide Coupling

An input fiber position where guided light can be checked is searched for while monitoring the exit end of the waveguide with an observation optical system. When the guided light can be checked at the exit end, the observation optical system is retracted. The positions of the waveguide and output fiber are adjusted until output light from the output fiber is obtained. Fiber alignment is made for the input and output fiber alternately until a maximum output power is obtained while monitoring the output power from the output fiber.

Where waveguide array and fiber array are aligned, the alignment is so performed that the coupling efficiency between the fiber at the left end and the waveguide is equal to the coupling efficiency between the fiber at the right end and the waveguide and that a maximum output is obtained.

9.3.1.3 Fiber–Fiber Coupling (Fiber Collimators)

A mirror with a surface perpendicular to the fiber-holding portion is placed. This mirror is made movable in the direction of the optical axis. The position is determined according to the working distance of the collimator. The lens is aligned such that light reflected from the mirror is recoupled into the fiber. The interval between the fiber and lenses needs to be adjusted according to the position of the mirror in the direction of the optical axis (working distance). A circulator is used to guide light from the light source to the fiber or to guide light, which is reflected from the mirror and recoupled into the fiber, into a power meter.

When fibers are coupled using a pair of fiber collimators fabricated in this way, pointing error produced during fabrication of the fiber collimators is corrected by adjusting the angle between the collimators. Usually, one fiber collimator is fixed, and the angle of the other collimator is adjusted as shown in Figure 9.8.

9.3.2 Passive Type

A light-emitting or light-receiving device is placed in position on a metal pattern by soldering, with the metal pattern on an SiOB (Si optical bench)[4] as shown in Figure 9.9. Fibers are held in V-grooves formed in the SiOB by anisotropic etching. The fibers can be respectively placed in position without alignment. Solder bump bonding is a method available to place optical elements in position on a metal pattern and to hold them. This is a means for self-aligning optical elements in desired locations by the surface tension of molten solder

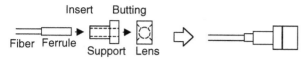

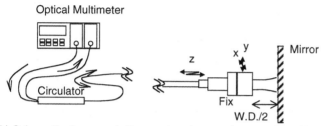

a) Schematic diagram of alignment for fiber collimator.

b) Schematic diagram of alignment assignment in assembly of fiber collimator.

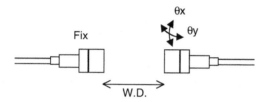

c) Schematic diagram of alignment assignment between a pair of fiber collimators.

Figure 9.8 Optical coupling technique of fiber-to-fiber using a pair of fibers.

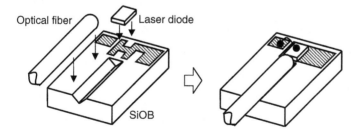

Figure 9.9 Schematic view of pigtail-type light source module using an SiOB.

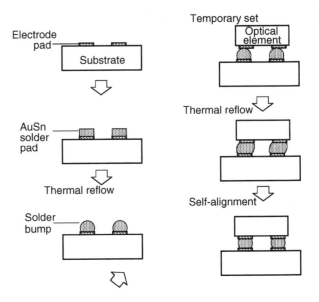

Figure 9.10   Self-alignment mechanism on the flip-chip bonding method.

and for holding the elements[5,6,7] as shown in Figure 9.10. An alignment accuracy of 2 to 3 μm is achieved by optimizing the size of the solder bumps. For an SiOB, V-grooves can be formed by etching and by selecting a crystal axis. Because the mask structure used for this etching is fabricated photolithographically, the V-groove and the metal pattern used to hold the element can be placed in position with high alignment accuracy, on the order of submicrons. The SiOB can be mass-produced by a wafer process, so a cost decrease owing to mass production can be expected.

9.4 TECHNIQUES FOR HOLDING OPTICAL FIBERS IN COUPLING OPTICS

Means for holding optical fibers can be roughly classified into four categories: adhesive, welding, soldering, and fusion splicing. These are different in size of equipment or machine. Furthermore, they are generally different in holding accuracy and throughput. It is necessary to use these means separately according to the required stability or cost.

9.4.1 Fixing Method Using Adhesive

It can be said that adhesive use is a low-cost fixing technique for small-quantity production because even dissimilar materials can be easily bonded together without expensive equipment. However, care must be exercised in applications for which long-term stability is required because the material is a resin. In addition, adhesive use is unsuited for applications in which high throughput is required because ultraviolet (UV)-curable adhesives need hardening time, as do room-temperature-setting adhesives and thermosetting adhesives.

9.4.1.1 Fiber-Ferrule

When an optical fiber is fixed to a ferrule, for which an optical fiber connector is a typical example, adhesive generally is used as shown in Figure 9.11. The outside diameter of the optical fiber and the diameter of the hole in the ferrule were accurately set, so the gap has a value of less than 1 μm. Therefore, the amount of positional shift caused by curing and shrinkage of the adhesive filling the space or expansion and shrinkage caused by temperature variations is also small. Because ZrO_2,

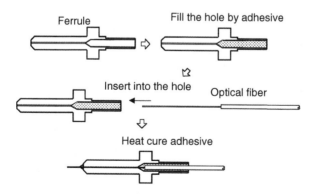

Figure 9.11 Schematic procedure of fixing optical fiber in ferrule using adhesive.

which does not transmit light, is often used as the material for ferrules, heat-curable epoxy adhesives are generally used. There are some considerations in selecting the adhesive. The adhesive needs sufficient hardness and waterproofness to withstand end face polishing. When a polarization-maintaining fiber is used, deterioration of the polarization-maintaining characteristics because of a decrease of the birefringence of the fiber must be prevented. For this purpose, it is important to select an adhesive and curing conditions that minimize the stress in the fiber.

9.4.1.2 Fiber-V-Groove (Array)

To fix optical fibers in arrayed V-grooves, a flat-plate structure is used from above to press the fibers into the grooves. Then, adhesive is used to fix the fibers as shown in Figure 9.12. The fibers are always placed in position within the V-grooves by the flat plate and thus are stably fixed. When the substrate

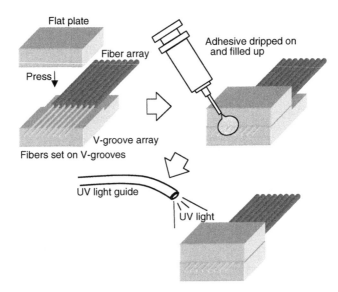

Figure 9.12 Fabrication of fiber array using ultraviolet (UV) curable adhesive.

with V-grooves is made of glass, the flat plate is made of the same material as the substrate. The fiber array is bonded with a UV-curable adhesive. When the substrate with V-grooves is made of Si, if the flat plate is also made of Si a UV-curable resin cannot be used. In the case of the substrate with V-grooves made of Si, the flat plate is often made of Pyrex glass, which is close to Si in coefficient of thermal expansion.

9.4.1.3 Fiber—Waveguide

As mentioned in the section about the methods of coupling, when an optical fiber and an optical waveguide are bonded together, the adhesive area is increased to secure sufficient adhesive strength. For this purpose, the fiber is held in a holey block or bead. The polished end faces are abutted against each other. Then, they are adhesively bonded. In the case of a multicore waveguide, a block with V-grooves is used. With respect to the waveguide, the end face is polished after loading a block having the same material as the substrate on the end face of the waveguide to make the bonding structure symmetrical (Figure 9.13).

In the case of a quartz-based waveguide, an adhesive having a refractive index that is coincident with that of quartz after curing is used. This suppresses the reflection loss at the bonding portion. Also, axial misalignment occurs during curing

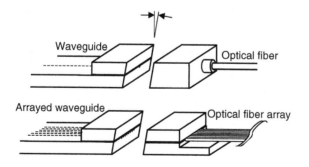

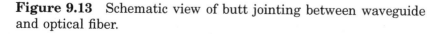

Figure 9.13 Schematic view of butt jointing between waveguide and optical fiber.

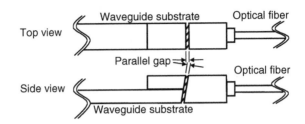

Figure 9.14 Schematic diagram of parallel gap for joining waveguide and fiber.

and shrinkage of the adhesive unless the bonded end faces are parallel (Figure 9.14). This would increase the excess loss.

9.4.1.4 Fiber-Lens

Fiber collimators are often used. Such a fiber collimator is made from parts of the fiber inserted in a ferrule and of a rod lens (typified by a GRIN lens) or drum lens. The drum lens is held within a sleeve with an adhesive.[8] The drum lens is fabricated by grinding out a ball lens. The lens-fiber eccentricity of the fiber collimator fabricated in this way is suppressed from about 10 μm to tens of micrometers by enhancing the accuracy at which the components are machined. If an eccentricity at this level occurs, however, an unallowable loss will be caused only if the outer references are formed on the fiber collimator and placed opposite each other. In this case, accordingly the fiber-collimator angle is adjusted to correct the eccentricity in the stage where the fiber collimator is mounted (Figure 9.15).

9.4.2 YAG Laser Welding

The greatest merit of the use of welding is that stability is obtained because members can be coupled at a high joining strength without dissimilar materials. Furthermore, the welding operation can be performed instantly, producing high throughput, which is another great merit. On the other hand, large equipment, such as an expensive laser machine, is necessary. Where laser spot welding is performed, a positional

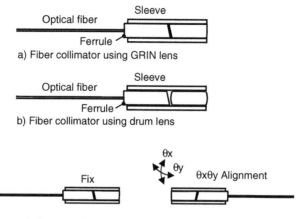

a) Fiber collimator using GRIN lens

b) Fiber collimator using drum lens

c) Alignment between optical fiber collimators

Figure 9.15 Typical fiber collimator using GRIN lens or drum lens.

deviation occurs between the members because of a series of state changes (i.e., temperature rise, expansion, melting, solidification, cooling, and shrinkage) at the welded location (Figure 9.16). Generally, members of identical material are welded at three points at the same time to reduce position deviation.[9] In addition, tilt between the joint surfaces of the members is a great factor causing postweld shift during welding.

Welding is done at three points simultaneously with YAG laser units uniformly spaced from each other by 120°. The members are further welded at three points after rotating the members through 60 or 180°. If necessary, the three-point welding is repeated to complement the spaces between the welded spots as shown in Figure 9.17. In this way, the bonding strength is increased. To minimize the deviation occurring during welding, the laser power is reduced in the initial welding phase, and provisional bonding is done. When a certain level of bonding strength is secured, the power is increased, and then the members are fastened nonprovisionally.

To perform a joining operation using welding technology, each element, including fibers, needs to be held in metal members. A stainless steel ferrule in which a ZrO_2 capillary

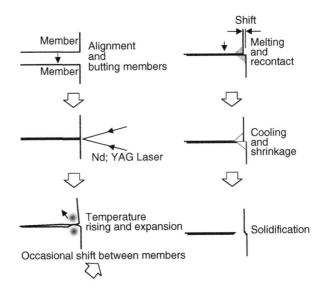

Figure 9.16 Mechanism of postweld shift between the welded members.

is mounted with a press fit is used as a fiber holder. Adhesive is used to hold the ferrule and fiber only. Generally, no adhesive is used for other purposes. Therefore, an aspheric lens molded in a metallic barrel is often used.

9.4.3 Fixing by Soldering

To fix glass elements such as fibers and lenses by soldering, the side surfaces of the elements to be fixed need to be metallized. To accomplish a hermetic structure using no adhesive to suppress aging of the optical elements within the package, the metallization is used in combination with solder sealing in some cases. In a proposed system, if an axial misalignment occurs during fixing, solder reflow, realignment of the fiber, and fixing are repeated to improve the assembly yield.

9.4.4 Fusion Splicing

Fusion splicing is a well-known technique for joining optical fibers. In particular, the end faces of fibers are cleaved regardless

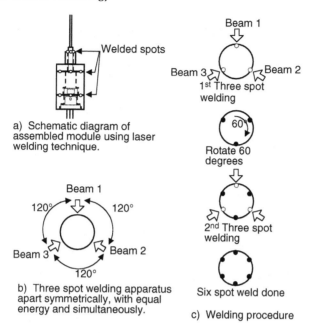

a) Schematic diagram of assembled module using laser welding technique.

b) Three spot welding apparatus apart symmetrically, with equal energy and simultaneously.

c) Welding procedure

Figure 9.17 Schematic procedure to minimize the postwelded shift in the Nd:YAG laser welding technique.

of whether they are single or tape fiber. The end faces are placed opposite and close to each other. They are melted and spliced together by an arc discharge. The fibers are roughly brought into coincidence with fiber clad as a reference. When the fibers are melted and solidify, self-aligning action works owing to the surface tension (Figure 9.18). A connection loss of less than 0.1 dB is normally achieved between similar fibers. In some examples, a quartz waveguide and an optical fiber were directly spliced together with a CO_2 laser (Figure 9.19). It is expected that this method also yields high stability because members are spliced together without using dissimilar materials.

Using this technique, a GI MMF is spliced to a single-mode fiber. The MMF is cut exactly to the length of a 0.25 pitch. Thus, this device can be used as a collimator lens.[10] In the fiber collimator fabricated in this way, no antireflective coating is necessary between the lens and fiber. Furthermore, any members for splicing are not necessary. Consequently, the lens and

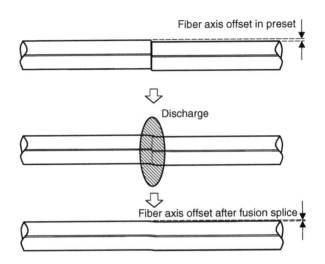

Figure 9.18   Decrease of fiber axis offset by self-aligning effect on fusion splice.

fiber made of the same material are fusion spliced, which will lead to cost savings. Also, high durability is obtained. However, normal GI MMFs result in collimator lenses with a short focal length. Therefore, in applications for which large working distances are necessary, the loss is increased. Hence, this type of collimator lens cannot be applied to these applications, a limitation of this process.

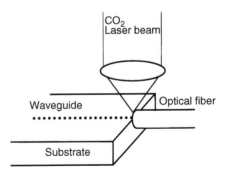

Figure 9.19   Schematic view of direct splicing waveguide and optical fiber with a CO_2 laser.

In some reported examples, a 0.4-mm GRIN lens was fabricated by a process similar to that of the GI MMF. The GRIN lens was fusion spliced to a single-mode fiber to fabricate a fiber collimator. The focal length of the lens was increased by increasing the diameter of the GI MMF. An average opposite insertion loss of less than 0.2 dB was accomplished at a working distance of 3 mm. In this example, the heat capacity was increased by increasing the lens diameter. A CO_2 laser producing a larger amount of energy was used for the fusion splicing.

9.5 INSPECTION TECHNIQUES FOR OPTICAL FIBER JISSO TECHNOLOGY

9.5.1 Measurements and Evaluations of Coupling Losses (Factors of Coupling Losses)

A pigtail module is assumed regarding measurements and evaluations of coupling losses. The inspection of the performance of fiber collimators is discussed. The performance especially creates factors leading to deterioration of the coupling efficiency.

9.5.1.1 Wavefront Aberration of a Collimator Lens

In a system in which light going out of an input fiber is again collected and coupled into an output fiber by an opposite collimator lens, the wavefront aberration in the collimator lens destroys the beam profile at the focal point, thus producing a factor leading to deterioration of the coupling efficiency to the output fiber. Therefore, it is necessary to use a collimator lens completely free from the aberration. The wavefront aberration in a collimator lens can be evaluated by interference fringes produced by causing reference and measured wavefronts to interfere with each other as shown in Figure 9.20. In considering the coupling efficiency to a single-mode fiber, the collimator lens is required to have a wavefront aberration of less than root mean square 0.035 because the relation between the wavefront aberration and the coupling efficiency is such

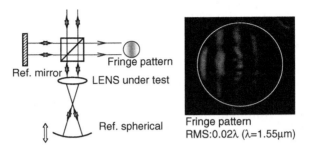

Figure 9.20 Schematic diagram of measurement wavefront aberration of collimator lens.

that the two collimator lenses are between the input and output fibers.

$$\eta > 1 - (2\pi/\lambda)^2[RMS]^2 \qquad (9.3)$$

Fiber collimator lenses have tended to have smaller diameters. With this trend, an interferometer for small-size lenses is prepared as the interferometer for measurement of wavefront aberrations.

9.5.1.2 Pointing Error

When a module is assembled by mounting fiber collimators (each fabricated separately) opposite each other, a pointing error produced on assembly of the collimators needs to be corrected. Accordingly, it is necessary to check that the pointing error of each fiber collimator alone is within a prescribed value. The pointing error can be inspected by a system that monitors the exit angle of the collimated light with an autocollimator adjusted and fixed relative to a reference on the outside of the fiber collimator. It is also possible to perform the inspection using a master collimator or mirror placed in a position relative to a reference on the outside of the fiber collimator. That is, the coupling efficiency to the master collimator or the recoupling efficiency of light reflected from the mirror into itself is measured, and then a check is done whether the obtained coupling efficiency is in excess of a prescribed value.

9.5.1.3 Diameter and Position of Beam Waist

The beam waist diameter varies depending on the focal length of the collimator lens or on the MFD of the single-mode fiber. If the beam waist diameter differs between the opposite collimators, a coupling loss occurs because of mode field mismatch. A beam profiler is used to measure the beam waist diameter.

The beam waist position after exiting from the output collimator is optimized according to the working distance that is necessary by the length of the element inserted between the collimators. In particular, the reflecting mirror is placed perpendicular to the fiber collimator. Variations in the recoupling efficiency of the reflected light into the fiber are measured with respect to the amount of movement of the reflecting mirror in the direction of the optical axis. The position of the reflecting mirror at which a maximum value is obtained is taken as the beam waist position (Figure 9.21).

9.5.2 Return Loss

After the optical fiber is mounted in a module, reflected light is produced because of the difference in refractive index at the surface of the elements and at the boundary

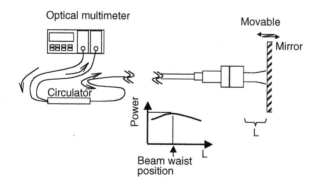

Figure 9.21 Schematic diagram of measurement beam waist position.

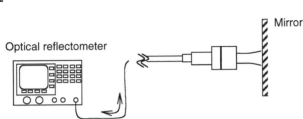

a) Schematic diagram of measurement to resolve multiple reflections spatially.

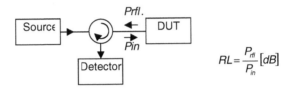

b) Principles of return loss measurements.

Figure 9.22   Return loss measurement techniques.

between the elements. The reflected light is recoupled into the fiber and becomes returning light. The returning light makes the wavelength and output power of the light source unstable. Multiple interferences vary the signal light intensity, so it is necessary to suppress the returning light. One method of measuring the return loss consists of measuring the amount of light returning from the whole object under investigation. Another method consists of measuring the position at which a return loss occurs and the amount of reflected light at that position (Figure 9.22). The methods are used separately according to the intended purpose. When the total length is hundreds of millimeters, such as in a module in which fiber is mounted, the position of the reflection point and the amount of returning, reflecting light can be measured using a reflectometer that uses optical low-coherence reflectometry (OLCR). In this case, the positional resolution at which the reflection point is detected is about 25 μm, and the sensitivity at which the intensity of the reflected light is detected is about 80 dB/m.

9.5.3   Evaluation of Temperature Characteristics

The completed module has the possibility that various characteristics are affected by temperature variations because of the solid-state stability of each element. For evaluation, the following method is adopted: Only the module of the measured system is placed in a constant-temperature oven. The temperature of the module is varied while monitoring the value to be measured. At this time, care should be exercised such that those portions (e.g., connectors) affected by the temperature variations are placed outside the oven.

9.6   OPTICAL FIBER ARRAYS

It is necessary to array optical fibers accurately in coupling the fibers to a light-receiving or light-emitting device or waveguide or coupling together optical fiber arrays. The required array accuracy differs according to whether the used optical fibers are multimode or single mode. In the case of single-mode fibers, the accuracy is 0.5 to 1 μm. In the case of MMFs, the accuracy is several micrometers. Generally, optical fiber arrays are one-dimensional (1D). According to the application, two-dimensional (2D) fiber arrays may be used.

In this chapter, optical fiber arraying means that it is indispensable for parallel optical interconnection, as described briefly. The Jisso technologies of fiber array using a V-grooved substrate are described in detail, with the array used frequently.

9.6.1   Fiber Arrays Using Substrates with V-Grooves

A substrate with V-grooves is used as a component for accurately arraying fibers because this is a relatively simple structure and because fiber arrays with high array accuracy can be fabricated. The types of substrate with V-grooves with different materials and different pitches are introduced here. Also, means for holding fibers in substrates with V-grooves are introduced.

TABLE 9.1   Material Features for Fiber Array V-Groove Substrate

Material	Ultraviolet transparency	Process	CTE ($\infty10^{-6}$)	V-angle
Silica		Grinding	0.5–0.6	Arbitrary
Pyrex glass		Grinding	3.3	Arbitrary
ZrO_2	X	Grinding	11	Arbitrary
Si	X	Etching	4.2	70.6°
Low-softening point glasses		Press molding	4.9–11.4	Arbitrary
Plastic (epoxy resin)	X	Injection molding	16	Arbitrary

9.6.1.1 Materials and Fabrication Methods

Fiber arrays using a substrate with V-grooves are often used because materials (listed in Table 9.1) can be selected from a relatively wide range of choices. The use of quartz glass makes it possible to reduce greatly the temperature dependence of the array accuracy. When such a fiber array is used for direct coupling to a waveguide, the temperature dependence of the coupling efficiency is suppressed by making it equal with the substrate material of the waveguide in coefficient of thermal expansion. (Generally, the same material is used.) This can also lead to improvement of durability. Selecting a material means selecting a coefficient of thermal expansion.

Transparency to UV light can be secured by using various types of glasses as substrate materials. This permits the use of a UV-curable resin in a later process step. This also needs to be taken into consideration when a substrate material is selected.

A substrate with V-grooves made of Si can be fabricated by etching wafers. Glasses with low softening points make it possible to mold a substrate with V-grooves from glasses. These Si low-softening-point glasses and plastics (epoxy resins) can be advantageously used for cost savings by mass production effects. Furthermore, when these materials are used, complex shapes can be obtained without involving large cost increases, thus providing wide latitude.

9.6.1.2 Core Pitch

The core pitch of the most general fiber arrays is 250 μm. Tape fibers arrayed at this pitch are also fabricated. General fiber clad diameter is 125 μm. Fiber arrays with a minimized pitch (known as half-pitch) of 127 μm are also fabricated. In this case, it is undesirable to rearray tape fibers with an array pitch of 250 μm into a pitch of 127 μm because bending of the fibers creates problematic loss variations. Therefore, the generally adopted method consists of stacking tape fibers in two stages and shifting the fibers laterally by an amount equal to the half-pitch such that the tape fibers in the upper stage are placed between the cores of the tape fibers in the lower stage. In this geometry, little bending of the fibers occurs (Figure 9.23).

9.6.1.3 Fixing Means

Adhesives are generally used to fix a fiber array to a substrate with V-grooves because this is a relatively simple method and because fibers can be held in V-grooves accurately. However,

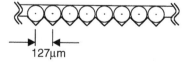

a-1) Half-pitch fiber array aligned

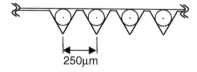

b-1) 250μm-pitch fiber array aligned

127μm

a-2) Cross section of fiber array
aligned in V-grooves with half-pitch.

250μm

b-2) Cross section of fiber array
aligned in V-grooves with 250-μm pitch.

Figure 9.23 Comparison of different structures for pitch of fiber array.

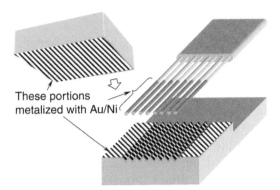

Figure 9.24 Metallized portions of the parts of a fiber array for soldering.

adhesives have limited heat resistance. In addition, they cannot be used in a hermetic package used to secure long-term reliability. Accordingly, when long-term reliability is required, soldering is used. To solder fibers made of glass to a substrate with V-grooves made of Si, glass, or ceramic, each element needs to be metallized. For this purpose, Au/Ni is sputtered on the elements, or the elements are coated with plating (Figure 9.24).

9.6.1.4 Method of Measuring Core Pitch

The core pitch of a fiber array is measured as follows: Light from a halogen light source is introduced into the core from the other end. The near-field pattern of the light at the exit end face is picked up by an observation system. The position of the center of the optical fiber is found by image processing. The center position is found from the amount of eccentricity with respect to the design position as shown in Figure 9.25. For this reason, measurement error might occur because of roughness or contamination of the observed end face. Fiber arrays with accurately polished end faces can be measured with a repeatability of 0.05 μm.

9.6.2 Combination with Microlens Arrays

To couple light going out of one fiber array into an opposite fiber array after causing the light to travel through a space,

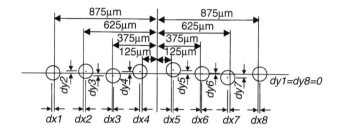

Figure 9.25 Schematic diagram of eccentricity in 250-μm pitch arrayed fiber.

the light going out of the fibers must be collimated using microlenses to improve the efficiency. A microlens array in which microlenses are arrayed accurately in the same way as in a fiber array is shown in Figure 9.26.

9.6.2.1 Types of Microlens Arrays

Microlens arrays include graded index lenses fabricated by ion exchange of glass[11]; lenses obtained by transferring a resist pattern that exhibits a spherical shape by resist reflow

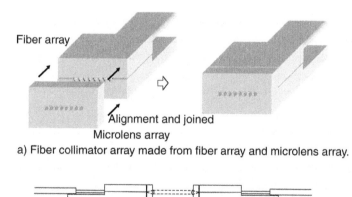

a) Fiber collimator array made from fiber array and microlens array.

b) Schematic view of coupling optics in a pair of fiber collimator arrays.

Figure 9.26 Schematic view of coupling optics in a pair of fiber collimator arrays.

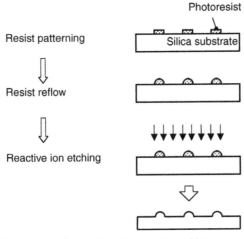

a) Fabrication of graded index microlens by ion diffusion.

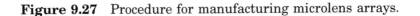

b) Fabrication of refractive microlens by etching.

Figure 9.27   Procedure for manufacturing microlens arrays.

onto a substrate material by etching[12] (Figure 9.27); lenses
fabricated by dripping a trace amount of solution like resin
that is volume controlled onto a substrate so that the resin
assumes a spherical form by surface tension and then curing
the resin; and lenses fabricated by manufacturing a mold with
a precision processing machine and molding plastic or glass
into a lens array. In any case, a lens array can be formed at
a pitch with submicron accuracy. A coupling efficiency of about

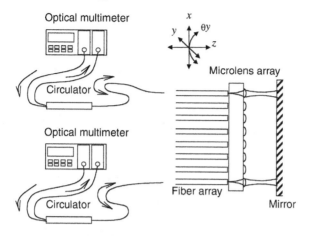

Figure 9.28 Schematic diagram of alignment process between fiber array and microlens array.

–0.5 dB can be accomplished from a single-mode fiber to a single-mode fiber by placing two lenses opposite each other.

9.6.2.2 Alignment between Fiber Arrays and Microlens Arrays and Their Fixing

Alignment between an array of fibers and an array of lenses is based on the assumption that the elements in each array are accurately arrayed. The alignment is carried out by aligning the cores of the lens and fiber at the opposite ends of the arrays as shown in Figure 9.28. In this case, alignment is performed to maximize the recoupling efficiency at the two cores at the opposite ends of the fiber array while monitoring the amount of the light recoupled into the fibers because of reflection of the collimated light by a mirror. The lens array is fixed by previously determining the substrate thickness such that the focal position of the lens agrees with the rear surface of the lens substrate, abutting the end face of the fiber array and the rear surface of the lens substrate against each other and then adhesively bonding the surfaces. In another method, a lens array substrate is molded. Holes for guide pins are formed simultaneously in the substrate. Alignment with

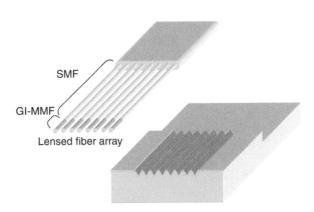

Figure 9.29 Another type of fiber collimator array using lensed fiber array and V-groove array.

a fiber array using an mechanically transferable (MT) ferrule can be performed without core alignment.

9.6.2.3 Lensed Fiber Plus V-Grooved Substrate

A lensed fiber obtained by processing the tip of an optical fiber can be directly aligned on a substrate with V-grooves and used as a fiber collimator array as shown in Figure 9.29. In this case, the core alignment accuracy is determined by the eccentricity produced during manufacture of the lensed fiber. In a lensed fiber fabricated by fusion splicing a GI MMF to a single-mode fiber, the fibers are equal in fiber-clad diameter. Consequently, the amount of eccentricity has a value of less than 1 μm because of centripetal action during the fusion splicing.

9.7 PACKAGING TECHNIQUES

9.7.1 Hermetic Structure

In applications for which high durability is required, the module is hermetically sealed to prevent the internal elements of the module from being dewed; otherwise, they would become corroded, and their optical characteristics would deteriorate. In the case of hermetic structure, the air inside the module is replaced by a dry, inert gas such as nitrogen, and circulation with the outside air is cut off. Methods of sealing include solder, glass, and resin sealing. Resins show moisture permeability under long-term, high-temperature, high-moisture conditions, so it is difficult to secure long-term reliability.[3] Solder sealing is widely used in hermetic seals because parts of metals such as Covar can be joined and because parts of glass typified by optical fibers can also be joined using metal coating. The portion of a modular package through which an optical fiber passes is sealed by metallizing the fiber clad from which the intermediate coat has been stripped and filling solder between this portion and the metal package (Figure 9.30).

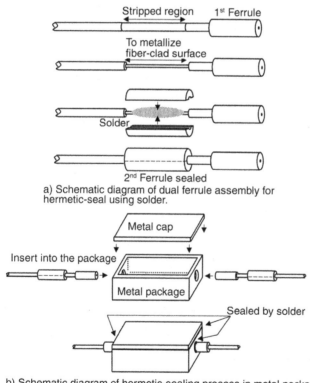

a) Schematic diagram of dual ferrule assembly for hermetic-seal using solder.

b) Schematic diagram of hermetic-sealing process in metal package by solder.

Figure 9.30   One hermetic sealing technique.

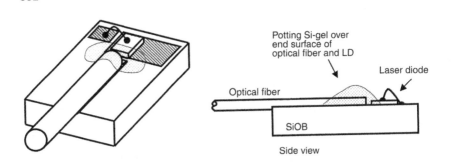

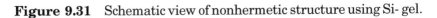

Figure 9.31   Schematic view of nonhermetic structure using Si- gel.

9.7.2   Nonhermetic Structure

Modules have been required to be fabricated at lower costs. Therefore, there is a movement toward reassessing hermetic structures. That is, the whole package is not hermetically sealed, but silicon gel is applied as a potting material to the surface of each optical element to prevent the surface from being dewed, thus protecting it.[13] The silicon gel is filled between the optical element and the fiber by a direct coupling structure (Figure 9.31).

9.8 CONCLUSION

The costs of an optical module are broken down into two categories: the cost of the elements and the cost of the Jisso process, including inspections. It is said that the cost of the Jisso process is twice as high as the cost of the elements for the following reasons: First, sophisticated Jisso technology is required. In addition, the specification differs among individual users, although the amount of production is not so large. That is, the manufacturer must cope with each individual user. Accordingly, a common platform is established to enhance the mass production effect. Utilization of optical modules is promoted to achieve lower costs. Furthermore, when in-line inspection that performs inspection as well as fabrication is introduced, the inspection cost is reduced. This also contributes to a decrease in the total cost of the module. However, the greatest obstacle in reducing the costs of mounting elements is to shift optical axis alignment, which has heretofore required active alignment, to passive alignment. It can be said that full automation of Jisso steps using passive alignment is the key to drastically reducing the costs of optical modules.

REFERENCES

1. P.O. Haugsjaa, Packaging of optoelectronic components, *CLEO-2001*, SC122 (2001).

2. N. Mekada, Practical method of waveguide-to-fiber connection: direct preparation of waveguide endface by cutting machine and reinforcement using ruby beads, *Appl. Opt.*, 29, 5096 (1990).

3. K. Shigihara et al., Modal field transforming fiber between dissimilar waveguides, *J. Appl. Phys.*, 60, 4293 (1986).

4. K.E. Bean, Anisotropic etching of silicon, *IEEE Trans.*, ED-25, 1185 (1978).

5. J. Sasaki et al., Self-aligned assembly technology for optical devices using AuSn bumps flip-chip bonding, *LEOS' 92*, 260 (1992).

6. T. Hayashi, An innovative bonding technique for optical chips using solder bumps that eliminate chip positioning adjustments, *IEEE Trans. CHMT*, 15 (1992).

7. M. Itoh et al., Use of AuSn solder bumps in three-dimensional passive aligned packaging of LD/PD arrays on Si optical benches, *Proc. 46th ECTC*, 1 (1996).

8. J. Fujikawa, Reliable mechanical splice using precision glass capillary for future single-mode fiber system, *Proc. 2nd Int. Symp. New Glass*, 110 (1989).

9. W.H. Cheng et al., Reduction of post-weld shift in semiconductor laser packaging, *Opt. Quantum Electron.*, 28, 1741 (1996).

10. W. Emkey and C. Jack, Analysis and evaluation of graded-index fiber-lenses, *J. Lightwave Technol.*, 5, 1156 (1987).

11. M. Oikawa et al., High numerical aperture planar microlens with swelled structure, *Appl. Opt.*, 29, No. 28, 4077 (1990).

12. A. Kouchiyama et al., Optical recording using high numerical-aperture microlens by plasma etching, *Jpn. J. Appl. Phys.*, 40, 1792 (2001).

13. I.P. Hall, Non-hermetic Encapsulation and Assembly Techniques for Optoelectronics Applications, *10th Eur. Microelectron. Conf.*, 117 (1995)

10

Optical Connectors

TAKEHIRO HAYASHI

CONTENTS

10.1   INTRODUCTION

Optical fiber communication networks have been developed rapidly and have become the most popular means of communication instead of a copper-based network. To accelerate the progress, an optical connector is the key device for progress. Many types of connectors are provided for proper selection. For example, SC, LC, and MU are for single glass fiber applications; MT-RJ, MPO, and MPX are for multiple glass fiber applications; F05 and F07 are for plastic optical fiber (POF) applications, and so on. In addition, new connectors have been developed for more advanced applications to increase the user-friendliness of the optical fiber interface and accelerate the development of the optical network.

In this chapter, the basic theories of optical connector design will be described, then the actual optical connectors will be introduced.

10.2 BASIC OPTICAL CONNECTOR

10.2.1 Applicable Optical Fibers and Their Characteristics

The features of some optical fibers that affect the design and performance of optical connectors are introduced. For details, please refer to other chapters in this book.

10.2.1.1 Glass Optical Fibers

Glass optical fibers are made of pure silica glass and have low attenuation and large bandwidth. Therefore, glass optical fibers can be applied to a wide range of transmission applications, such as tens of meters of a local-area network (LAN) or thousands of kilometers of long-distance transmission systems with data transmission speeds of hundreds of gigabits per second. Further, even 1-Tb/sec bandwidth can be achieved using wavelength division multiplexing (WDM) technologies.[1] There are two typical types of optical fibers from the fiber parameter and performance points of view: single-mode (SM) optical fiber and multimode (MM) graded index (GI) optical fiber (Figure 10.1). There are many subcategories in SM and MM optical fibers, depending on the application, (e.g., polarization-maintaining

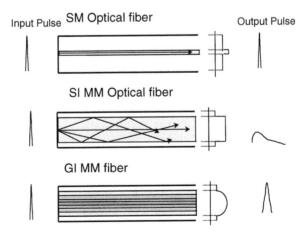

Figure 10.1 Optical fiber structures.

and absorption-reducing [PM or PANDA] optical fiber, dispersion-shifted [DS] optical fiber, nonzero dispersion-shifted [NZ] optical fiber, etc.), the core size (mode field diameter is the correct description) of a SM optical fiber is about 9 μm, and the core size of a MM GI fiber is 50 μ or 62.5 μm generally. Because of the small core size, micron or submicron accuracy of core alignment must be achieved so that the optical loss at the connecting point is low. As a result, precision components are required for connecting glass optical fibers.

10.2.1.2 Plastic Optical Fibers

POFs are made of acrylate polymer and have a MM step index (SI) structure (Figure 10.1). Although the attenuation is quite high and the bandwidth is small (e.g., attenuation and bandwidth of a typical POF are 100 dB/km or more and several megahertz per kilometer, respectively), because the physical core size is rather large (about 1 mm), handling and connection are easy, and large-tolerance plastic components can be accepted for connecting POFs. This feature enables provision of low-cost components and cable assemblies. Coupled with the electromagnetic noise-free feature, POFs have been widely introduced in automotive, home electronics, and industrial equipment applications with transmission distances of tens of meters or less.[2] In contrast, studies to expand the bandwidth and transmission distance have been made, and some new POFs with GI (or GI-like) structure have been developed and introduced.[3,4] Optical fibers realize a bandwidth of hundreds of megahertz per kilometer, and this would accelerate to introduce the new POFs to high-speed transmission applications.

10.2.1.3 Propagation Characteristics

The characteristics of optical fibers reflect the features of propagation states caused by the structure of optical fiber. For a SM optical fiber, launched light, if the light wavelength meets the SM transmission condition, it can obtain a stable SM propagation state immediately regardless of the launching condition. Consequently, optical performance at the connecting point, such as optical loss, can be measured with a high degree of reliability.

MM optical fibers show rather complex propagation characteristics compared to SM optical fibers. In other words, the values of optical measurements may fluctuate according to launching conditions, propagation distance, bending of fibers, and so on. Light propagating in an optical fiber eventually reaches the steady state of modal power distribution, which is determined by optical fiber structure. In the steady state, light can propagate stably with little change of modal power distribution.

Before steady state, the modal power distribution is in either an overfilled or underfilled state. In the overfilled state, high-order modes, including cladding propagation modes, exist in the optical fiber. Light in these modes is leaked to the outside gradually during propagation in the optical fiber, and the modal power distribution approaches steady state. In contrast, in the underfilled state, only low-order modes exist in the optical fiber. Modal power distribution gradually approaches steady state with modal conversion to higher order modes. Figure 10.2 shows an example of the overfilled and underfilled states. A light source with a larger launching angle (e.g., a light-emitting diode [LED]) than numerical aperture (NA) of the optical fiber creates the overfilled state, and a light source with a small launching angle (e.g., a laser diode [LD]) than the NA of the optical fiber creates the underfilled state. In either of these states, the light propagation condition changes easily because of a bending radius change caused by handling the optical fiber; therefore, a measurement reliability problem may be encountered. For

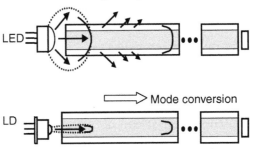

Figure 10.2 Overfilled and underfilled states.

reliable optical measurement, particularly for a MM optical fiber, reaching the steady state launching condition is a key issue. A mode scrambler is the most popular and effective solution to avoid the problem.

10.2.2 Key Factors of Insertion Loss at Connecting Point

Regarding optical fiber connectors, there are four major factors that generate optical fiber insertion loss (IL), which is power loss at a connecting point:

1. Optical fiber misalignment caused by the design/ structure.
2. Fresnel reflection caused by a gap between optical fiber facets.
3. The condition of the optical fiber facet.
4. Difference of optical fiber parameters.

In this section, the first three topics are discussed. Some practical factors that affect IL performance, such as dust and scratches of the optical fiber facet in the field, are also discussed.

10.2.2.1 Core Misalignment

IL caused by core misalignment is determined by the optical connector structure and its degree of accuracy, except for the accuracy of the optical fiber. In other words, IL can be estimated if errors of optical connector components, particularly components which align fibers, are known. Here, there are three important elements for misalignment: (1) lateral offset, (2) angular offset, and (3) gap (Figure 10.3).

Because there have been many studies about the relation between these elements and IL for SM and MM GI glass optical fibers, I introduce the descriptions in JIS (Japan Industrial Standards) C5962.[5] The relation between IL of SM optical fiber and the lateral and angular offset is given as

$$L = -4.34\left[\left(\frac{d}{\omega_0}\right)^2 + \left(\frac{\pi\bar{n}\omega_0\theta}{\lambda}\right)^2\right] \text{ (dB)} \qquad (10.1)$$

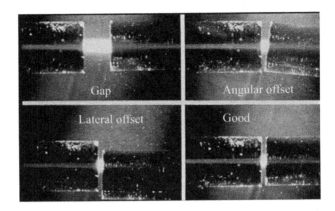

Figure 10.3   Three elements for misalignment.

where

$$\frac{\omega_0}{a} = 0.65 + \frac{1.619}{V^{\frac{3}{2}}} + \frac{2.879}{V^6} \qquad (10.2)$$

where d is lateral offset, θ is angular offset, $n = 1$, a is the core radius, λ is the wavelength, and V is the normalized frequency. The result for a SM fiber with a 10-μm core at a 1310-nm wavelength is shown in Figure 10.4.

The relation of a MM GI optical fiber at steady state is given as

$$L = -10\log_{10}\left[1 - \frac{u^4 D^2}{8(u^2 - 4)}\right] \text{ (dB)} \qquad (10.3)$$

where

$$D^2 = \left(\frac{d}{a}\right)^2 + \left[\frac{2\bar{n}\tan(\frac{\theta}{2})}{NA}\right]^2 \qquad (10.4)$$

where $u = 2.405$, d is the lateral offset, θ is the angular offset, $n = 1$, a is the core radius, and NA is the numerical aperture. The result for a MM GI fiber with a 50-μm core at a 850-nm wavelength and steady state is shown in Figure 10.5.

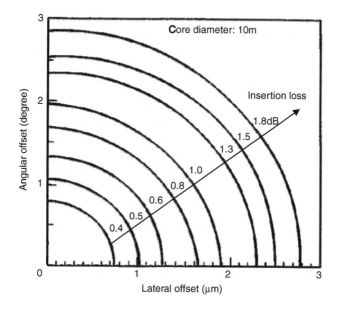

Figure 10.4 Relation between power loss and lateral/angular off-set for a SM fiber.

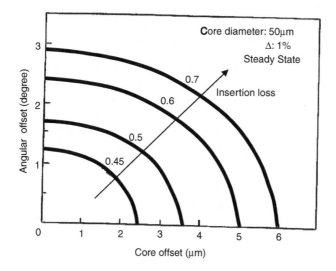

Figure 10.5 Relation between power loss and lateral/angular off-set for a MM GI fiber at steady state.

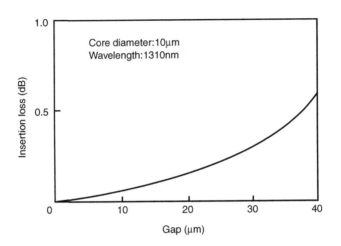

Figure 10.6   Relation between insertion loss and gap for a SM fiber.

The relations between IL and gap for SM and MM GI fibers are given in Equation 10.5 and Equation 10.7, respectively, and the results are shown in Figure 10.6 and Figure 10.7.

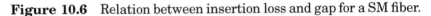

$$L = -10 \log_{10} \left[\frac{1 + 4Z^3}{(1 + 2Z^2)^2 + Z^2} \right] \text{ (dB)} \qquad (10.5)$$

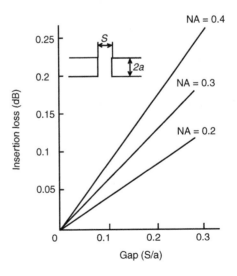

Figure 10.7   Relation between insertion loss and gap for a MM fiber.

where

$$Z = S/kn_2\omega_0 \tag{10.6}$$

where k is $2\pi/\lambda$, and n_2 is the refractive index of cladding.

$$L = -10\log_{10}\left(1 - \frac{S}{4a}NA\right) \text{ (dB)} \tag{10.7}$$

Here, S is the gap.

Because POF has not been used to any extent in high-performance applications, few practical studies have been performed. New POFs with performance equivalent to glass optical fiber have been developed, and the design of the connector has become an important issue. Figure 10.8 shows the results.[6] The chart shows the relation between IL and core offset of the SI POF and the GI POF where D is the core diameter and d is the core offset. IL of SI POF increases linearly with core offset. In contrast, the GI POF shows a modest parabolic increase, and the IL in the small offset area is lower than that for the SI POF.

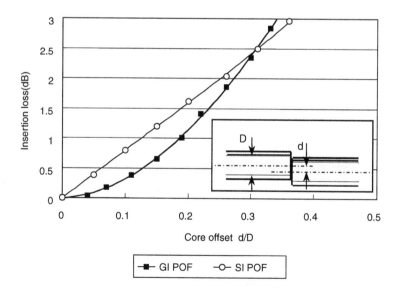

Figure 10.8 Insertion loss of POF caused by core offset.

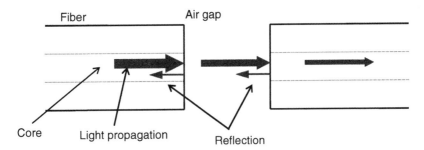

Figure 10.9   Fresnel reflection at connecting point.

10.2.2.2 Fresnel Reflection

Light reflects on passing the boundary of different refractive index materials. The reflection is known as the *Fresnel reflection*. At the connecting point of optical fibers, this is similar (Figure 10.9). The general Fresnel reflection ratio is given as

$$R = \frac{(n_1 \sin\alpha_1 - n_2 \sin\alpha_2)^2}{(n_1 \sin\alpha_1 + n_2 \sin\alpha_2)^2} \tag{10.8}$$

where n_1 and n_2 are refractive indexes of material, and α_1 and α_2 are the light launching angles.

For an optical fiber connection with an air gap,

$$\alpha_1 = \alpha_2 = 90° \tag{10.9}$$

Equation 10.8 is simplified to

$$R = \frac{(n_1 - n_2)^2}{(n_1 + n_2)^2} \tag{10.10}$$

Then,

$$L = 2 \times -10 \, \text{Log}_{10}(1 - R) \, (\text{dB}) \tag{10.11}$$

gives the IL caused by the reflection of the glass optical fiber. Because the light passes the boundary of glass and air twice, the result needs to be doubled. The refractive indexes of glass and air are $n_1 = 1.46–7$ and $n_2 = 1$, respectively. The result is 0.309 dB.

10.2.3 Optical Requirements

The most important performance of an optical connector is IL, especially for the glass optical fiber connectors used in a high-performance communications network. The attenuation of an optical fiber is approaching 0.1 dB/km in a SM fiber, so 0.5-dB IL at a connecting point makes the transmission distance about 4 to 5 km shorter. From the system designer's point of view, the IL at a connection point should be 0, if possible. Losses of 0.75 dB/connection for a LAN system and 0.5 dB/ connection for a telecommunications network system are recognized as industrial standards for glass optical fibers.

Regarding POFs, there are several different types of fibers with a variety of attenuations. The IL requirement is determined system by system, but generally a 2 dB/connection is practical.

As discussed in section 10.2.2, the IL is determined by the degree of the precision of the optical fiber alignment. Table 10.1 shows a guideline for the degree of precision to achieve the required IL for a glass optical fiber.

Another important performance is return loss (reflectance), which is defined as the ratio between launched optical power and reflected optical power (Figure 10.10). One reason why the return loss needs to be specified is, as explained regarding the Fresnel reflection, the reflection increases the IL. Another reason is that the reflected light propagates in

TABLE 10.1 Guideline for Alignment Precision

	Insertion loss			
	0.75		0.5	
Type of fiber	Single-mode	Multimode graded index (50)	Single-mode	Multimode graded index (50)
Core offset (μm)	1.5	6	1	3.5
Angular offset (°)	1.5	3	1	1.5
Gap (μm)	0	0	0	0

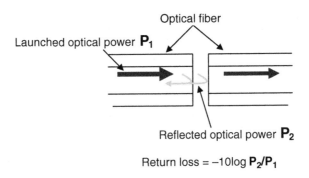

Return loss = $-10\log P_2/P_1$

Figure 10.10 Definition of return loss.

the fiber in the opposite direction, then the light enters the light source. When the light is highly coherent (e.g., a laser), the returned light interferes with light in the laser, and this degrades the coherency (Figure 10.11). Requirements are 25 dB return loss for a medium speed (100 Mb/sec to 1 Gb/sec) digital transmission system and 40 dB or more return loss for a high speed (over 2.5 Gb/sec) digital transmission system and analog (video signal).

10.2.4 Other Requirements

Depending on the application, specific mechanical and environmental performances are required. Major items are described in this section. The requirements are specified in industrial standards, and details are described in section 10.6.

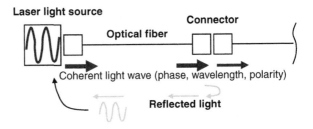

Figure 10.11 Influence of reflected light.

1. Mechanical performance.
   A. Intermatability.
   B. Coupling/cable retention.
   C. Durability/repeatability.
   D. Impact/shock.
   E. Vibration.

2. Environmental performance.
   A. Operating/storage temperature.
   B. Humidity.
   C. Corrosion resistance.
   D. Dust resistance.

10.2.5 Concepts for Precision Fiber Alignment

Many precision optical fiber alignment methods have been proposed to meet the optical requirements shown in section 10.2.3 as well as economical requirements. Here, the popular methods/concepts utilized in the industry will be explained.

10.2.5.1 Precision Capillary

The precision capillary is one of the simplest concepts. The simple glass or ceramics capillary has a precision through hole, and the diameter of the hole is slightly (about 1 μm) larger than the optical fiber. Because the standard diameter of a glass optical fiber is between 124 and 126 μm, a hole diameter between 125 and 127 μm is applicable. Glass fibers are cleaved with a cleaving tool to obtain a mirror surface (Figure 10.12), and the fibers are inserted and mated in the capillary. This concept is used for a permanent splice rather than a connector; however, so-called ferruleless connectors have been proposed.

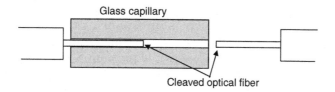

Glass capillary

Cleaved optical fiber

Figure 10.12 Capillary concept.

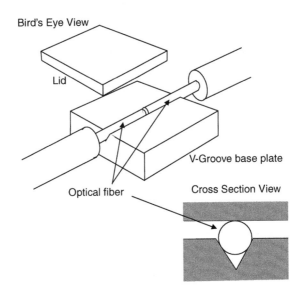

Figure 10.13   V-groove concept.

10.2.5.2 V-Groove

The V-groove concept is another simple way to align optical fibers (Figure 10.13). The advantage of the V-groove is that it can hold fibers without a radial gap between the fiber and the V-groove, so that the lateral offset is minimized. As for the precision capillary concept, because it is difficult to apply this concept to a connector, the V-groove is used for a permanent splice.

10.2.5.3 Combination of the Precision Capillary and Alignment Sleeve

Both the precision capillary and the V-groove concepts are good for precision optical fiber alignment, but it is hard to realize detachability, which is an important connector function. The combination of the precision capillary and alignment sleeve concepts provides a detachable function with minimal increase of alignment error (Figure 10.14). This is the most popular design for an optical fiber connector for a single fiber.

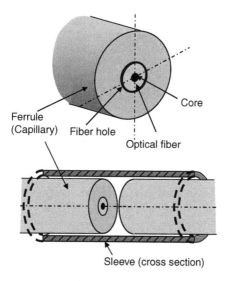

Figure 10.14 Capillary/sleeve concept.

10.2.5.4 Combination of the Precision Capillary and Alignment Pins

Alignment (or guide) pins instead of a sleeve are used to align precision capillaries (Figure 10.15). The concept is good for handling multiple fibers in a rectangle ferrule. The MT connector[7]

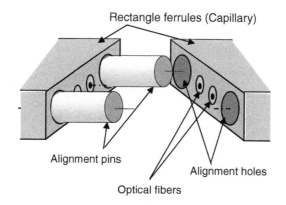

Figure 10.15 Capillary/alignment pin concept.

and its family are well-known applications of the concept. As for the other concepts, fibers are fixed by glue, and the end facet is polished.

10.3 BASIC DESIGNS FOR GLASS OPTICAL FIBER CONNECTORS

In the theoretical factors for IL, the study of optical connector design is performed to optimize all the factors to minimize geometric errors, except for uncontrollable optical fiber parameters. The points below are key to high-performance optical connector design.

1. Optical fiber core aliment structure.
2. End facet finish.
3. Physical fiber contact structure.

In addition, the following practical factors need to be considered to design a reliable and user-friendly connector:

4. Durability/repeatability for cycle connection.
5. Maintaining good connection.
6. Coupling strength.
7. Protection of connection.
8. Ease of handling.
9. Interface compatibility.

These points are discussed in this section.

10.3.1 Key Components

10.3.1.1 The Ferrule and Sleeve

A cylindrical ferrule is a standard optical fiber connector design (Figure 10.16). For glass optical fibers, optical fibers are firmly fixed by glue, and the end facet of the ferrule is polished to tens of nanometers of smoothness. A pair of polished ferrules is aligned in a sleeve with a slit. The inner diameter of the sleeve is slightly smaller than the diameter of the ferrule. The ferrule opens the sleeve so that the ferrule can enter the sleeve, and the sleeve generates force to hold the ferrule because of elasticity. Because of the force from the

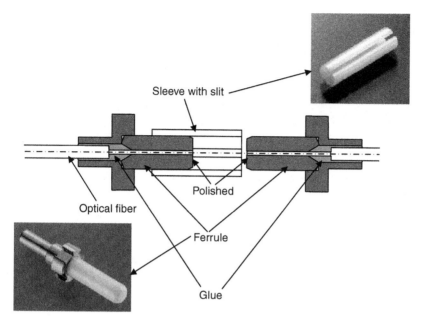

Figure 10.16 Ferrule/sleeve structure.

sleeve, the ferrules are held tightly in it. As a result, the ferrules can be aligned precisely. To avoid an air gap between the optical fibers, a PC (physical contact) polishing technique[8] to make the ferrule end spherical is utilized (Figure 10.17).

10.3.1.2 The Ferrule and Guide Pin

A design with a rectangular ferrule with two guide holes and multiple fiber holes is another standard for a multifiber

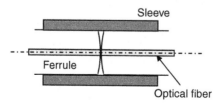

Figure 10.17 Dome PC.

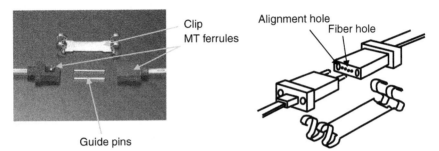

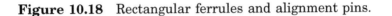

Figure 10.18   Rectangular ferrules and alignment pins.

optical connector (Figure 10.18). The fibers are fixed by glue, and the surface is polished to tens of nanometers of smoothness. The ferrules are aligned with precision pins. To avoid an air gap between the optical fibers, another PC polishing technique that makes the optical fibers protrude is utilized (Figure 10.19).

10.3.2   Materials

10.3.2.1   Ceramics

The majority of cylindrical ferrules are ceramic. In the past, alumina was widely used for cylindrical ferrules; however, alumina has a rather large grain boundary, which makes alumina bulk porous. In addition, alumina is a hard material, with a Mohs hardness of 9. Because of these physical properties,

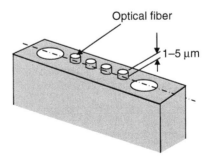

Figure 10.19   Protruded PC polish.

expensive diamond polishing material is required to polish alumina ferrules, and it is hard to control the recess depth of the optical fiber surface from the ferrule surface because of the hardness difference. Grains of alumina often drop out of the ferrule, and the dropped grains make scratches on the optical fiber. Diamond powder provides machining stress to glass during polishing, and this makes a layer with a high refractive index on the surface of the glass. This creates a reflection at the connector surface.

The zirconia ferrule was developed to solve these problems. The hardness of zirconia is close to that of silica glass, so inexpensive softer material such as silicone carbide can be used to polish the zirconia ferrule, and the fiber recess/protrusion can be controlled to the order of tens of nanometers. Because the stress applied to zirconia is absorbed by a martensitic transformation mechanism, this prevents grains from dropping out of the ferrule. Because of the softer polishing material, less stress is applied to the glass fiber, which improves the return loss performance significantly. In Table 10.2, the typical physical properties of alumina and zirconia are shown.

10.3.2.2 Polymers

Polymers, so-called superengineering plastics, are widely used for molding components with a few micrometers of dimensional error because of their excellent physical properties. Epoxy, PPS (polyphenylenesulfide), and LCP (liquid crystal

TABLE 10.2 Typical Physical Properties of Ceramics

	Alumina Al_2O_3	Zirconia ZrO_2
Bulk specific gravity (g/cm^3)	3.9	6
Water absorption (%)	0	0
Bending strength (MPa)	350	1100
Compression strength (MPa)	2900	3000
Young's modulus (GPa)	400	210
Poisson's ratio	0.24	0.32
Hardness (GPa)	15	13
Thermal expansion coefficient (room temperature ~200°C) (/°C)	6.5×10^{-6}	9.5×10^{-6}

TABLE 10.3 Typical Physical Properties of Engineering Plastics

	Epoxy	PPS	LCP
Bulk specific gravity (g/cm^3)	1.9	1.9	1.85
Water absorption (%)	0.01	0.01	—
Bending strength (MPa)	180	190	230
Bending modulus (MPa)	20,000	18,000	18,500
Shrinkage after molding (%)			
Flow direction	0.25	0.2	0.1
—Direction perpendicular to flow	0.25	0.5	0.35
Thermal expansion coefficient (°C)			
—Flow direction	1.5 × 10^{-6}	1 × 10^{-5}	1 × 10^{-5}
—Direction perpendicular to flow	1.5 × 10^{-6}	2 × 10^{-5}	4 × 10^{-5}

polymer) have excellent properties of after-molding shrinkage, thermal expansion coefficient, water absorption, and material strength (Table 10.3). Many studies to obtain the optimum compound/blend and molding conditions for these materials and progress in fine machining technologies for precision dies and molds have provided plastic ferrules with submicron precision. Compared with ceramic components, the degree of precision of molding components is slightly inferior; postprocesses such as grinding, honing, and polishing are not required except for the removal of flash, which makes the manufacturing cost lower.

10.3.3 Key Parameters and Factors

10.3.3.1 Cylindrical Ferrules

Cylindrical ferrules have 2.5- and 1.25-mm industrial standard size. The key parameters of the dimensions are shown in Table 10.4 and Table 10.5. As shown in the tables, the dimension tolerances are tightly controlled; in addition the hole size can be chosen to fit the ferrule tighter to the given fiber to minimize the gap between the fiber and the ferrule. Some different types of tip shapes are available for different applications. In contrast, because it is hard to specify slit sleeve by dimension, ferrule withdrawal force (Figure 10.20) is used to specify the performance. The range of ferrule withdrawal force

TABLE 10.4 Key Parameters for a 3.5-mm Cylindrical Ferrule

		Single-mode grade	Multimode grade
Outer diameter		2.499 ± 0.0005 mm	2.499 ± 0.001 mm
Hole diameter		0.124 0.125 ± 0.001 mm 0.126	0.126 0.128 ± 0.002 mm
Hole concentricity		0.0014 mm	0.004 mm
Tip shape		Flat(PC) Dome(PC) Conic(APC) Step(APC)	

TABLE 10.5 Key Parameters for a 1.25-mm Cylindrical Ferrule

		Single-mode grade	Multimode grade
Outer diameter		1.249 ± 0.0005 mm	1.249 ± 0.001 mm
Hole diameter		0.124 0.125 ± 0.00 mm 0.126	0.126 0.128 ± 0.002 mm
Hole concentricity		0.0014 mm	0.004 mm
Tip diameter		φ0.53 MU φ0.60 LC φ0.90 LC	

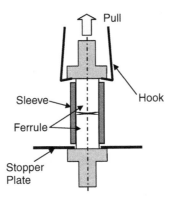

Figure 10.20   Ferrule withdrawal test method.

of the sleeves for 2.5-mm ferrules is between 2.0 and 5.9 N, and for 1.25-mm ferrules it is between 1.0 and 2.5 N. Details of the test method are described in JIS C 5961 7.4 or International Electrotechnical Commission (IEC) 1300-3-33.[9]

Generally, there are two major fabrication methods for ceramic cylindrical ferrules: extrusion and injection methods. The process outlines are shown in Figure 10.21 and Figure 10.22, respectively. Extrusion fits for high-volume production, and injection fits for high-mix low-volume production because complex shapes can be made using the injection molding method and less postprocessing is required.

Another important technology for fabrication is a precision measuring method. Because tolerance of the component is less than 1 μm, a normal conventional measuring machine cannot inspect it. Measurement of the hole concentricity is the biggest challenge. A new video measurement system that is a combination of high-resolution and high-sensitivity CCD (charge-coupled device) and high-precision XY stage has solved the issue. Figure 10.23 shows an example of the video measuring system.

10.3.3.2 Rectangular Ferrules

Rectangular multifiber ferrules are known as MTs and mini-MTs in the industry. The key parameters of the ferrule are

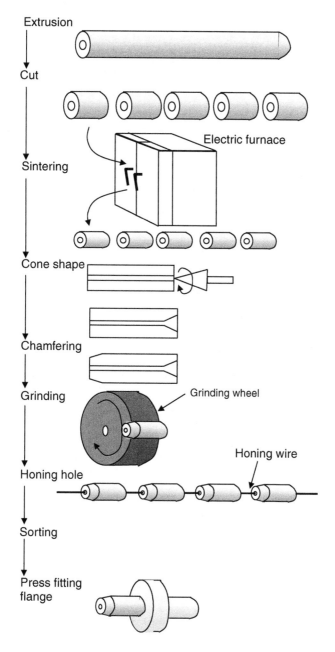

Figure 10.21 Ceramic cylindrical ferrule fabrication process (extrusion).

Injection

Cut

Electric furnace

Sintering

Chamfering

Grinding

Honing hole

Sorting

Press fitting flange

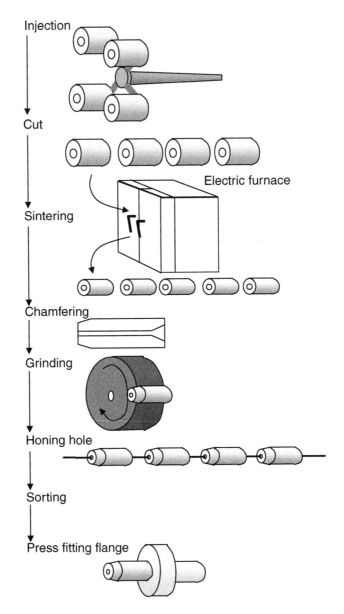

Figure 10.22 Ceramic cylindrical ferrule fabrication process (injection).

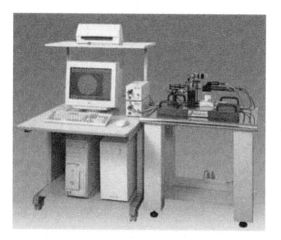

Figure 10.23 Example of ferrule concentricity measurement system. Photograph provided by Moritex Co., Tokyo, Japan.

shown in Table 10.6 and Table 10.7. MT/mini-MT ferrules are made using molding technologies, and the degree of precision required is unusual for polymer components. The progress of material, molding machines, and machining technologies to make dies and molds has been significant, and fine polymer optical components have been realized. The transfer molding

TABLE 10.6 Key Parameters for a MT Ferrule

		Single-mode	Multimode
Guide hole diameter		0.709 to 0.701 mm	0.699 to 0.701 mm
Guide hole pitch		4.597 to 4.603 mm	4.597 to 4.603 mm
Fiber hole diameter		126 to 127 μm	126 to 128 μm
Fiber hole position offset		1.5 μm	3 μm

TABLE 10.7   Key Parameters for a Mini-MT Ferrule

	Single-mode	Multimode
Guide hole diameter	+0.699 to 0.701 mm	0.699 to 0.701 mm
Guide hole pitch	+2.597 to 2.603 mm	2.597 to 2.603 mm
Fiber hole diameter	+126 to 127 μm	126 to 128 μm
Fiber hole position offset	+1.5 μm	3 μm

method for thermoset materials and injection molding for thermoplastic materials are major molding techniques for MT/mini-MT ferrules. In transfer molding, a material tablet that is a compound of epoxy, hardener, and mineral filler is preheated and melted; then, the melted tablet is put in the mold pot. The material is pressed by the plunger and injected into the cavities through the gates. Next, the material is cured in the hot mold. In the transfer molding process, because the material is liquefied, high injection pressure is not required. The material flows into cavities gently, which keeps the extra-fine pins in the cavities that form the holes of the MT ferrule from bending because of material pressure (Figure 10.24). The injection molding method formerly was used to make low-cost multimode ferrules because the shrinkage after molding is relatively larger, but the molding cycle time is much faster than for the transfer molding method. However, the liquidity and shrinkage of material such as PPS and LCP have improved, enabling production of high-performance SM ferrules.

Although inspection of hole sizes and positions is more difficult than for the single-fiber cylindrical ferrule, a similar video measurement system can provide a precise measurement (Figure 10.25).

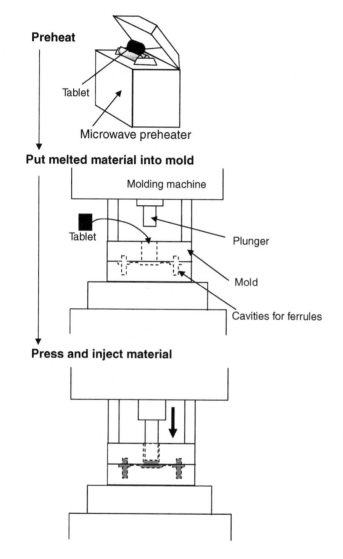

Figure 10.24 Transfer molding process.

10.3.3.3 Core Offset and Direction

It is generally thought that the core offset is defined by a combination of fiber core offset and ferrule hole offset. Although the fiber core offset and the ferrule hole offset are the major causes that affect the total core offset, the gap between the fiber and

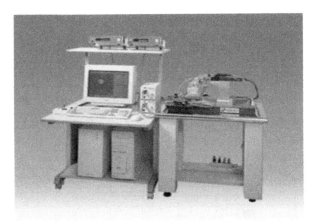

Figure 10.25   Example of a MT ferrule hole position measurement system. Photograph provided by Moritex Co., Tokyo, Japan.

the hole also contributes to the total core offset (Figure 10.27). Figure 10.26 shows that a combination of the gap between the fiber and the ferrule hole, the fiber core offset, and the ferrule offset, defines the total core offset. Different ferrule hole sizes are provided in the industry, so one can be chosen to best fit the given fiber to minimize the gap.

Although the gap is minimized by the selection of an optimum ferrule, a certain actual core offset still remains. If connecting such connectors at random in a rotating direction, the distance between the mated cores would be twice that of the

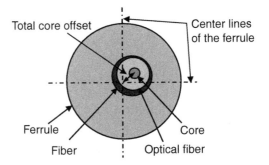

Figure 10.26   Influence of the gap between the hole and the fiber.

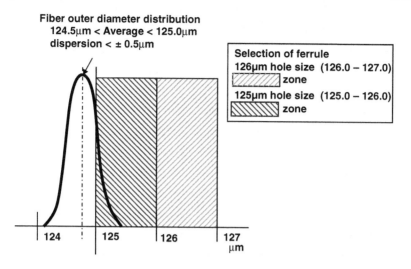

Figure 10.27 Estimation of the gap.

original core offset in the worst case. This means that random connection in a rotating direction makes the IL distribution worse. The method to tune the core offset direction using a connector keying function is provided. The actual ferrule has four key ways that decide ferrule orientation in the plug housing (Figure 10.28). Without a tuning direction, the cores are randomly in a certain area of the ferrule in the rotating direction (Figure 10.29). By orienting the core positions the key way, the cores concentrate in the smaller area (Figure 10.30). This makes the apparent core distance between the connectors much smaller. As a result, the IL distribution improves. Figure 10.31

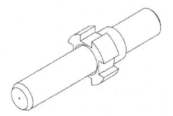

Figure 10.28 Actual ferrule.

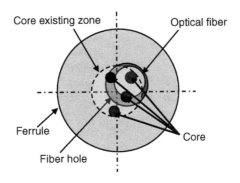

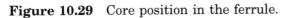

Figure 10.29   Core position in the ferrule.

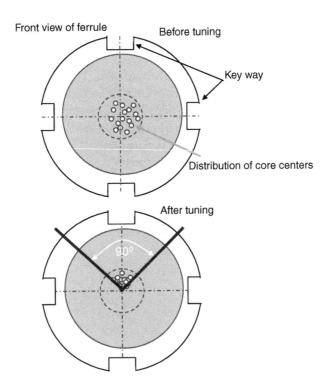

Figure 10.30   Core location before and after tuning.

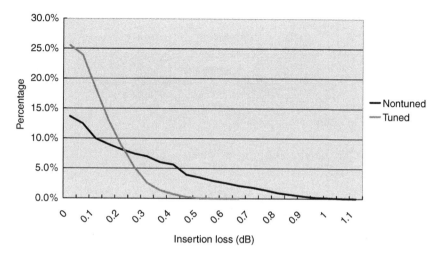

Figure 10.31 Insertion loss comparison, tuned vs. nontuned.

shows the difference of IL distribution for a tuned connector vs. a nontuned connector.

Note that the core tuning technique can be applied to a cylindrical ferrule with only the keying function.

10.3.3.4 Polishing Technologies for Low Reflection

The surface of a glass optical fiber connector is usually finished to a mirror surface by polishing. If an air gap exists between the fiber ends on the connecting fibers, the Fresnel reflection must occur. The PC connecting method with a dome-shape polishing technique was developed to realize a low-reflection (high return loss) connection (Figure 10.17). The dome shape is produced by polishing the ferrule ends using a polishing film (such as diamond, alumina, or silicone carbide film) on an elastic base (Figure 10.32). After the dome-shaping process, the fiber is polished two or more steps to obtain a smooth mirror surface using softer and smaller polishing material. Figure 10.34 shows a typical polishing machine.

The PC polishing method provides a return loss of about 30 dB. Because of the requirements for a connector with less

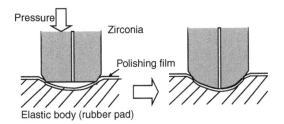

Figure 10.32   Dome-shaping process.

reflection for a high-quality SM transmission system, angled PC[10] and advanced PC[11] were developed. As described, polishing glass with a hard material creates an affected layer on the surface of the glass, and the refractive index of the layer is slightly higher than that of the original glass. When such connectors are connected, reflection occurs at the boundary of the affected layer and the original glass even though the glass surfaces physically contact each other. Angled PC polishes at a given angle (usually 8 or 9°) and prevents propagation of reflected light in the core because an angled PC surface releases reflection from the core (Figure 10.33). Advanced PC polishes with soft polishing material (such as silica) to remove the affected layer, reduces reflection. The return loss performances of angled PC and advanced PC are about 55 and 45 dB, respectively.

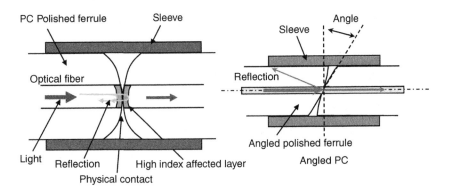

Figure 10.33   Reflection at affected layer and effect of angled polish.

Figure 10.34 Example of a polishing machine. Photograph provided by Seikoh Giken Co. Ltd., Chiba, Matsudo, Japan.

PC using the dome-polishing technique can be applied to cylindrical ferrules. In contrast, the protrusion polishing technique for PC is used for rectangular (MT/mini-MT) ferrules. Because the ferrules are made of polymer, protrusion is obtained easily with soft polishing material such as silica or cerium oxide.

10.3.3.5 End Face Geometries

For securing a good connection, the geometry of the polished surface needs to be specified. For cylindrical ferrules, radius of curvature, dome offset, and fiber withdrawal/protrusion are specified in an international standard. The definitions of these items and specifications are shown in Figure 10.35.

Although end face geometry specifications for rectangular ferrules are under discussion in the IEC, the key dimensions and the tolerances are shown in Figure 10.36.

Because tens of nanometers of accuracy are required, a special inspection machine that uses an optical interferometric effect is required to measure end face geometries. Figure 10.37 shows an example of the interferometric inspection machine.

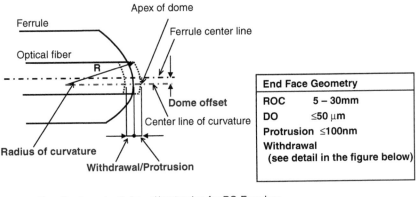

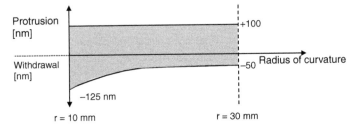

Specification of withdrawal/protrusion for PC-Ferrules:

Figure 10.35 Definitions and specifications of end face geometry for cylindrical ferrule.

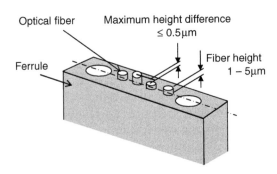

Figure 10.36 Definitions and specifications of end face geometry for rectangular ferrule.

Figure 10.37 Example of end face geometry inspection machine. Photograph provided by NTT Advanced Technologies Co., Tokyo, Mitaka, Japan.

10.3.4   Other Mechanical Functions and Physical Designs for Optical Connectors

A primitive connection is completed by mating a pair of ferrules using a sleeve or alignment pins for the cylindrical ferrule and the rectangular ferrule, respectively. However, again optical connectors that work in practical applications need to be considered for the following performances:

1. Durability/repeatability for cycle connection.
2. Maintaining good connection.
3. Coupling strength.
4. Connection protection.
5. Ease of handling.
6. Interface compatibility.

To fulfill performance requirements, optical connectors must have many other functions. The following mechanisms and functions are the major concerns for the practical applications:

1. Floating structure.
2. Coupling function.

3. Guiding function/connection sequence.
4. Intermatability.

These are discussed in this section.

10.3.4.1 Floating Structure

A combination of a pair of ferrules and a sleeve (for a cylindrical ferrule) or alignment pins (for a rectangular ferrule) could work as a complete connector in a still environment. However, the actual place to install optical connectors would be a dynamic environment from mechanical and thermal points of view. To eliminate the influence from the environment, optical connectors need to have a shell structure that is generally called *housing* to protect ferrules and sleeve or alignment pins. In addition, not to carry the force or impact applied to the connector to the connecting point, the ferrules and sleeve/pins have to be independent from the housing. Figure 10.38 and Figure 10.39 show cross sections of typical optical connectors for the cylindrical and rectangular ferrules, respectively. It is shown that ferrules are suspended by a spring in the housing and mechanically independent of the housing.

10.3.4.2 Coupling Mechanism

A reliable coupling mechanism is required to maintain a good connection. A screw coupling mechanism formerly was popular.

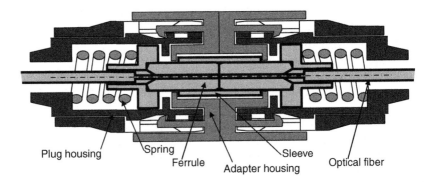

Plug housing   Spring   Ferrule   Sleeve   Adapter housing   Optical fiber

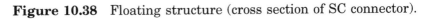

Figure 10.38   Floating structure (cross section of SC connector).

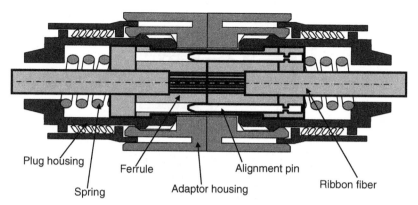

Plug housing   Ferrule   Alignment pin

Spring   Adaptor housing   Ribbon fiber

Figure 10.39   Floating structure (cross section of MPO connector).

FC is the world's first high-performance optical connector employing a screw mechanism and was developed by NTT (Nippon Telegraph and Telephone) in the late 1970s.[12] Although the screw mechanism is reliable and intuitive for handling, the connecting process with a rotating screw was difficult for installers. This was replaced by an easier mechanism. The most popular mechanisms are the push–pull coupling mechanism employed by SC, MU,[13] MPO,[14] LIGHTLAY MPX™ (Tycoelectronics Corporation Harrisburg, PA),[15] and so on and the single-lever mechanism, which is similar to the RJ-45-type plug/jack employed by LC™ (OFS BrightWave LLC, Norcross, GA), MT-RJ. These connectors are introduced in section 10.6.

10.3.4.3 Guiding Function and Connection Sequence

For easy handling during connection/disconnection as well as protection of the ferrule end face, the guiding function and connection sequence are important. Figure 10.40 shows an example of the process of connection. On starting a connection, the plug housing needs to contact the adaptor housing before the ferrule contacts to the entrance of the sleeve (or the sleeve holder, a part of the adapter housing) because if the order is reversed the ferrule contacts first; this may occur when the ferrule end hits the adapter housing and make scratches on the

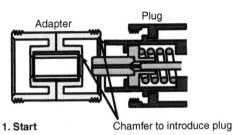

1. Start   Chamfer to introduce plug

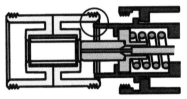

2. Adapter housing start to guide plug housing.

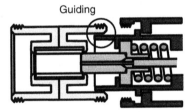

3. Ferrule comes into sleeve.

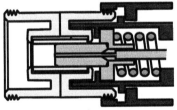

4. Connector plug bottoms

Figure 10.40   Guiding function and connection sequence.

glass fiber surface. Next, the mated plug housing and adapter housing position the ferrule around the center of the sleeve, and the ferrule is introduced smoothly into the sleeve, led by the chamfer at the end of the sleeve holder (Figure 10.40). At this time, if guiding the plug housing by the adaptor housing is poor, the ferrule surface may hit the sleeve holder. A combination of an adequate sequence of connection and guidance provides easy and error-proof operation.

10.3.4.4 Intermatability

The last important issue is compatibility among suppliers. For the good component availability, multisourcing in the market is essential for popularization of an optical network. In this case, however, connection of connectors manufactured by different suppliers always occurs. If the detail dimensions of a mating part are not determined, a problem of inability to make a connection might occur. To avoid this problem, industrial standards specify the interface standard to ensure intermatability. Industrial standards are established by major national standardization organizations such as JIS and EIA/TIA as well as international standardization organization such as IEC. Figure 10.41 shows an example of an interface standard by JIS. In the interface standard, two data planes are important: mechanical reference and optical reference. The mechanical

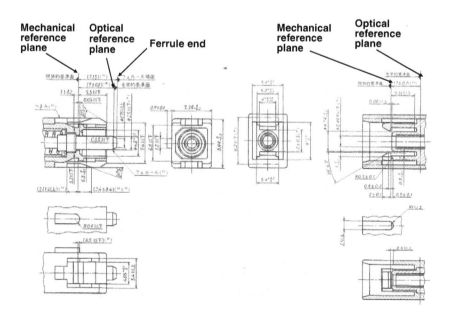

Figure 10.41 Interface standard of SC (from JIS C 5973).

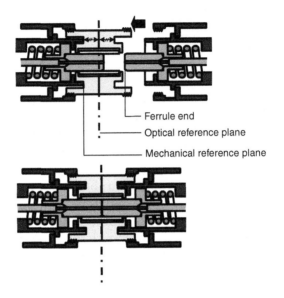

Figure 10.42   Reference planes.

reference plane is the position that physically contacts the mechanical reference plane of the opposite adapter (or plug). The optical reference plane is the point at which optical fibers are designed to meet. Because actual connectors have fluctuations of spring pressures, ferrule lengths, and holding force of the sleeve, fibers do not always meet at the point; however the data are important for connector design (Figure 10.42).

10.4 BASIC DESIGNS FOR PLASTIC OPTICAL FIBER CONNECTORS

Regarding POFs, the combination of a cylindrical ferrule and sleeve is also a popular structure. Because of the large core size and ease of alignment, the method to fix the fiber and finish the end facet are simple. Piercing to hold the jacket and fusion and flattening the finish of the end facet are standard for terminating a connector on a POF (Figure 10.43).

Because the components (ferrule and sleeve) of POFs can accept hundredths of millimeters of tolerance because of the

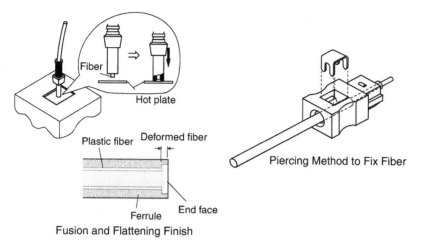

Figure 10.43 General termination method for POF.

large core, more general plastic materials such as PBT (poly-butyleneterephthalate) and PA (polyamide) can be applicable.

The performance requirement of IL of POF cable assemblies is about 2 dB, and return loss is not taken into account, so PC technology or floating design is not required. The key to designing a POF connector is cost minimization. A simple one-piece structure with a simple friction lock or single-lever lock mechanism is common (Figure 10.44).

10.5 TEST METHODS

The performance of optical connectors and cable assemblies have to be certified that they are in compliance with requirements or specifications. The performances to test are not only optical performance such as IL and return loss but also mechanical and environmental performance.

10.5.1 Optical Performance Tests

Because it is impossible to specify optical performance of optical connectors themselves, the optical performances are specified for the cable assemblies. Here, IL and return loss are major items. Because optical measurements for glass optical

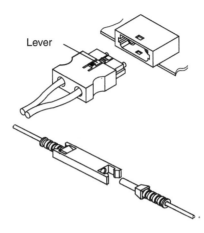

Figure 10.44 Lever lock and friction lock.

fiber cable assemblies are complex, detailed procedures are specified to minimize measurement error.

10.5.1.1 Insertion Loss

A basic IL test method is shown in Figure 10.45. Light launching condition is the most important parameter for reliable measurement, especially for MM glass optical fibers. As described, because measurement error caused by modal optical power distribution is significant, the mode condition needs to be kept at the steady state. In the case of a 50/125 GI MM fiber, a

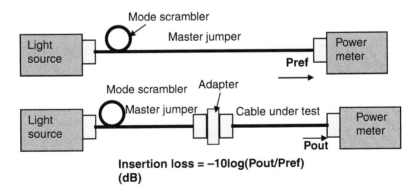

Insertion loss = −10log(Pout/Pref)
(dB)

Figure 10.45 An example of insertion loss test method.

TABLE 10.8   Single-Mode Master Jumper

Core offset	≤0.5 μm
Launching angle	≤0.2°
Tolerance of ferrule diameter	≤0.5 μm

launching condition with 0.195 ± 0.01 NA generates the steady state. A mode scrambler, which is optical fiber wound on a cylinder with a certain diameter, can provide the steady state. For a SM optical fiber, mode control is much easier than for a MM optical fiber because light other than for the propagation mode immediately leaves the optical fiber. Optical fiber several meters after the light source is enough to reach the SM condition.

In contrast, master jumpers are important for the SM because small core offset affects the IL performance greatly. Master jumpers for SMs need to meet the requirements in Table 10.8.

10.5.1.2 Return Loss

Return loss is measured as shown in Figure 10.46. Using the combination of splitter and total reflection mirror, the initial

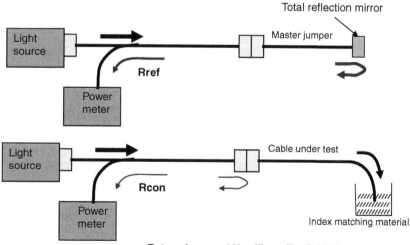

Figure 10.46   An example of return loss test method.

input power is measured. After that, the total reflection mirror is replaced by the cable for the test. Then, the reflection that occurs at the connector is measured. Return loss performance is important for SM PC, advanced PC, and angled PC connectors.

10.5.2 Other Performance Requirements and Test Conditions

Other than optical performances, optical connectors are required for mechanical performances and environmental performances. The major items are indicated in Table 10.9.[16]

10.6 CONNECTORS AND APPLICATIONS

10.6.1 Standard Connectors

Popular optical connectors already standardized in the industry are introduced in Table 10.10, Table 10.11, and Table 10.12.

10.6.2 Connectors for Special Applications

10.6.2.1 Backplane Connectors

Communication transmission equipment that uses many optical fibers have been introduced with the development of the broadband IP (Internet Protocol) network. In the equipment, optical components or circuits on daughter cards are connected to optical cables or circuits on the backboard. In such applications, the connectors must have a plug-in (so-called blind mate) function. MU and LIGHTLAY MPX™ are designed to fit a backplane application and have a smart guiding mechanism to connect each other and absorb the position errors from the board design. Figure 10.47 and Figure 10.48 show examples of backplane applications.

10.6.2.2 Ultrahigh-Density Connectors

For the solution of ultrahigh-density and ultrawideband data transmission, parallel optics has been studied. Two methods are proposed. One is a stacked fiber array using a conventional

TABLE 10.9 Mechanical and Environmental Requirements and Tests

Item	Requirements (single-mode)	Test method	Test condition
	Mechanical requirements and tests		
Vibration	Insertion loss after test ≤0.7 dB No damage, deformation, or crack	JIS C 5961 7.1 or IEC 61300-2-1	Frequency 10 to 55 Hz Amplitude 0.75 mm Direction X–Y–Z Duration 1 Hr/direction
Shock	Insertion loss after test ≤0.7 dB No damage, deformation, or crack	JIS C 5961 7.2 or IEC 61300-2-9	Peak acceleration 981 m/s^2 (100 g) Duration 6 msec Frequency 5
Mating durability	Insertion loss after test ≤0.7 dB Connection without fail	JIS C 5961 7.3 or IEC 61300-2-2	Frequency 500
Tensile strength of coupling mechanism	Insertion loss after test ≤0.7 dB No damage, deformation, or crack	JIS C 5961 7.4 or IEC 61300-2-6	Applied force 68.6 N
Cable retention	Insertion loss after test ≤0.7 dB No damage, deformation, or crack	JIS C 5961 7.11 or IEC 61300-2-4	Maximum pulling force 50 N Maximum pulling speed 50 mm/min

Environmental requirements and tests

Salt mist	No corrosion	JIS C 5961 8.1 or IEC 61300-2-26	Duration 48 H Concentration of salt solution 5%
Temperature cycle	Insertion loss after test ≤0.7 dB No damage, deformation, or crack	JIS C 5961 8.2 or IEC 61300-2-22	Temperature −25°C to 70°C Duration of plateau 30 min Ramp of temperature 3°C/min Frequency 2
Composite temperature/ humidity cycle	Insertion loss after test ≤0.7 dB No damage, deformation, or crack	JIS C 5961 8.4 or IEC 61300-2-21	Temperature −10°C to 65°C Relative humidity 95% Duration of plateau 3 H Ramp of temperature <1°C/min Cycle time 24 H Frequency 10

TABLE 10.10 Single Glass Optical Fiber Connectors

Type	Ferrule size, material	Coupling mechanism	Applicable fiber type	Typical IL	Typical RL	Application
FC	φ2.5 mm zirconia	Screw	SM	0.15 dB	30 dB PC 45 dB Ad PC 55 dB APC	Test equipment
			MM	<0.1 dB	NA	
SC	φ2.5 mm zirconia	Push-pull	SM	0.15 dB	30 dB PC 45 dB Ad PC 55 dB APC	Transmission equipment, Patch panel
			MM	<0.1 dB	NA	
LC	φ1.25 mm zirconia	Lever	SM	0.15 dB	30 dB PC 45 dB Ad PC NA	Network equipment, Patch panel
			MM	<0.1 dB	NA	
MU	φ1.25 mm zirconia	Push-pull	SM	0.15 dB	30 dB PC 45 dB Ad PC NA	Transmission equipment, Patch panel, Back plane
			MM	<0.1 dB		

Note: AdPC, advanced physical contact; APC, angled physical contact.

TABLE 10.11 Multiglass Optical Fiber Connectors

Type	Ferrule size, material	Coupling mechanism	Applicable fiber type	Typical IL	Typical RL	Application
MT-RJ	2.5 × 4.5 mm polymer	Lever	SM MM	0.3 dB 0.15 dB	45 dB PC NA	Network equipment, Patch panel
MPO	2.5 × 6.5 mm polymer	Push-pull	SM MM	0.3 dB (0.15 dB low loss) 0.15 dB	55 dB APC NA	Transmission equipment, Patch panel
MPX	2.5 × 6.5 mm polymer	Push-pull	SM MM	0.3 dB (0.15 dB low loss) 0.15 dB	55 dB APC NA	Transmission equipment, Patch panel Back plane

Note: RL, return loss.

TABLE 10.12 Plastic Optical Fiber Connectors

Type		Ferrule size, material	Coupling mechanism	Applicable fiber type	Typical IL	Application
F-05		2.5 mm Polymer	Friction	ϕ1 mm SI POF	2 dB	Digital audio
F-07		2.5 mm Polymer	Lever	ϕ1 mm SI POF	2 dB	Data transmission equipment

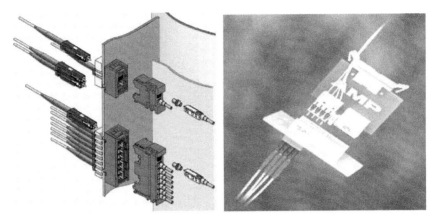

Figure 10.47   MU backplane applications.

MT ferrule; the other is a next-generation ferruleless connecting solution that is proposed as a fiber PC and BF (bare fiber) connector by NTT Laboratories.

10.6.2.2.1 MPO with Stacked Fiber Array (Para-Optix)

Para-Optix™ is designed based on MPO using a ferrule with a matrix of fibers instead of conventional single lines of fibers. The maximum fiber count is 72. Figure 10.49 shows the interface of Para-Optix and the transceiver.

Figure 10.48   LIGHTLAY MPX high-density backplane application installed with electrical connector.

Figure 10.49   Ultrahigh-density connecting solution (Para-Optix).

10.6.2.2.2 Ferruleless Connector (Fiber PC/BF Connector)

Fiber PC is designed for further high-density connection of the future pure optical cross connect and switch in which thousands of fibers are installed. As the name indicates, this connector does not use any type of ferrule. The fiber itself works as a ferrule and a spring. Bare optical fibers are inserted in a microhole and aligned; in addition, a buckling fiber provides a force to press the fiber surface forward.[17,18] Figure 10.50 and Figure 10.51 show the application of fiber PC. Fiber PC would be a breakthrough for high-density connection technology.

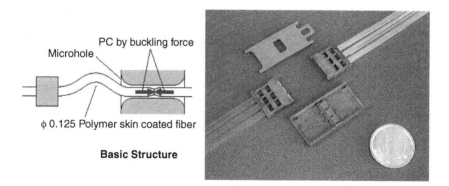

Figure 10.50   Concept of fiber PC and 36-fiber fiber PC connector. Photographs provided by NTT Photonics Laboratories, Atsagi, Kanagawa, Japan.

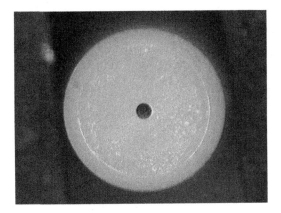

Figure 10.51   Contaminated ferrule surface.

10.7 GENERAL CONSIDERATIONS FOR HANDLING OPTICAL CONNECTORS

As described, optical connectors show good performance for keeping the core alignment accurate; in contrast, the operators always should handle optical connectors carefully so they demonstrate the best performance. From this point of view, small microns of contamination on the glass surface and scratches on the surface from careless handling would affect optical performance. The safety of operators is another important issue. The optical power transmitting in the fiber increases because of the improvement of light source and light-coupling technologies. Also, the light is invisible because the light wavelength for the high-power application is 1310 or 1550 nm. Therefore, operators may be unaware that they are seeing a high-power light.

General concerns and considerations for handling optical connectors are described in this section.

10.7.1 Cleaning

Microdust and contamination on the optical connector surface affect optical performance (Figure 10.51). However, optical connectors cannot always be handled in a clean room. An appropriate cleaning method for the surface of the connector

Figure 10.52 An example of a cleaning kit. Photograph provided by NTT Advanced Technologies Co.

is important. A cotton swab moistened with alcohol was a former method, but this method caused the remaining alcohol to make an adhesive thin film on the fiber, and wiping with a used swab might recontaminate the surface. For the cleaning solution, special cleaning kits that use a microfiber cloth are provided (Figure 10.52). The microfiber has a 3- to 5-μm diameter, and the cloth has the capability to trap small particles in the mesh of the fibers.

10.7.2 Scratching Fibers

Return loss performance has been improved drastically with the improvement of polishing technologies. As a result, even a small scratch (called a *white scratch*) cannot be allowed on the core region, especially for such applications as an analog transmission system and optical fiber amplifiers, for which high return loss is required. For PC polish, the optical fiber surface is the apex of the connector; therefore, the connector without a protection cap may hit something hard and scratch the fiber. To prevent this from happening (1) caps should always be on plugs and (2) the surface should be checked by a 200× microscope before connecting connectors.

10.7.3   Eye Safety

The light wavelengths used in the optical transmission system are 1310 and 1550 nm for a SM fiber. These wavelengths are not visible; therefore, operators may accidentally watch the connector when it is in service, and light may be launched into the eyes. The total optical power is about tens or hundreds of milliwatts even for high-power applications. The density of power is significant. Such a high-power light may damage human eyes, which may cause vision loss in the worst case. Eye safety protection such as safety glasses should be worn, and operators should always maintain attention. Connectors with a shutter mechanism are proposed, which would be a better solution.

REFERENCES

1. M. Koga et al., Present technologies and expectations for wavelength-division multiplexing transmission systems, *J. IEICE*, 83, 569 (2000).

2. D. Seidl et al., Application of POFs in data links of mobile systems, *Proc. 7th Int. POF Conf.*, 205 (1998).

3. T. Ishigure et al., High bandwidth GI POF and mode analysis, *Proc. 7th Int. POF Conf.*, 33 (1998).

4. M. Naritomi et al., Advanced perfluorinated GI-POF applications, *Proc. 10th Int. POF Conf.*, 201 (2001).

5. Japan Industrial Standard Committee, *Generic Rules of Connectors for Optical Fiber Cables*, JIS C 5962, Japanese Standards Association, Tokyo, Japan, p. 16, 2001.

6. N. Sudo et al., Optical connector for high performance plastic optical fiber, *Proc. 8th Int. POF Conf.*, 213 (1999).

7. S. Nagasawa et al., Mechanically transferable single mode multi fiber connector, *IOOC Tech. Dig.*, 4, 48 (1989).

8. N. Suzuki et al., Low insertion loss and high return loss optical connector with spherical convex-polished end, *Electron. Lett.*, 22, 110–112 (1986).

9. Japan Industrial Standard Committee, *Test Methods of Connectors for Optical Fiber Cables*, JIS C 5961, Japanese Standards Association, Tokyo, Japan p. 42, 1997.

10. M. Takahashi, Study of characteristics and fabrication method of connector for single mode optical fiber, list of summaries of doctoral theses, Shizuoka University, 1997, http://www.lib. shizuoka.ac.jp/gakui/gde/gde162.htm.

11. T. Saitoh et al., Polishing technology for very high return loss optical fiber connector, *IEIEC EMC90-23*, 90, 23 (1990).

12. N. Suzuki et al., A new demountable connector developed for a trial optical transmission system, *IOOC Tech. Dig.*, B-10-4, 351 (1977).

13. S. Iwano et al., Compact and self-retentive multiferrule optical backplane connector, *IEEE J. Lightwave Technol.*, 10, 1356 (1992).

14. S. Nagasawa et al., A single-mode multifiber push-on type connector with low insertion and high return losses, *ECOC*, MoB1-7, 49 (1991).

15. T. Hayashi et al., LIGHTLAY MPX™ optical backplane interconnection system, *IEICE EMD2000-16*, 100, 7 (2000).

16. Japan Industrial Standard Committee, *Test Methods of Connectors for Optical Fiber Cables*, JIS C 5961, Japanese Standards Association, Tokyo, Japan, p. 39, 1997.

17. M. Kobayashi et al., Study of fiber management for optical board integration, *Trans. IEICE*, J84, 774 (2001).

18. A. Ohki et al., Development of 60Gb/sec-class parallel optical interconnection module (ParaBIT-1), *IEICE Trans. Electron.*, E84-C, 3, 295–303, (2001).

11

Parallel Optical Links

SHINJI NISHIMURA and HIROAKI NISHI

CONTENTS

11.1 OVERVIEW

In this chapter, the following six points are described.

1. Concept of parallel optical links.
2. Design of parallel optical link devices.
3. Parallel optical link devices.
4. Local-area network (LAN) standard for 10Gb Ethernet.[1]
5. Wide-area network (WAN) OC-192 and OC-768 standards.[2]
6. Server input/output (I/O) standard for InfiniBand.[3]

The interface standard and the multisource agreement (MSA) for the package are described in the last three topics.

11.2 CONCEPT OF PARALLEL OPTICAL LINKS

High-performance and high-density logical large-scale integrated (LSI) technologies have rapidly accelerated the development of such information-processing equipment as large-scale computers, routers, and disk storage equipment.[4-9] However, the high-speed wiring devices that connect this equipment have become performance bottlenecks. To overcome these bottlenecks over short distances of up to several kilometers, the parallel optical link is a possible candidate.

Optical fiber has excellent electromagnetic interference (EMI) characteristics for high-speed and low-loss (long transmission length) data transmission. These characteristics make

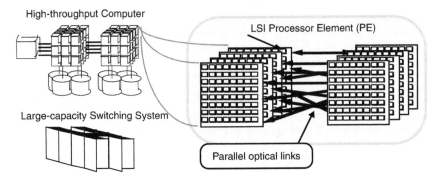

High-throughput Computer

LSI Processor Element (PE)

Large-capacity Switching System

Parallel optical links

Figure 11.1 Application area for parallel optical links between processor elements.

the best use of large capacity and the high-density interconnection between information processing equipment.[4–9]

Figure 11.1 is an image of a computer system that uses parallel optical links. Parallel optical links connect the devices and boards easily with optical fibers. Optical/electrical (O/E) and electrical/optical (E/O) conversion modules are set up in the vicinity of the logical LSI. As a result, optical modules replace bipolar LSIs with complementary metal-oxide semiconductor (CMOS)-LSIs, and the use of optical links becomes possible through the higher speed logical LSI I/O ports.

Figure 11.2 is a basic composition of the parallel optical link. Data transmission is achieved using O/E and E/O conversion modules that integrate the array of two or more channels. Large-throughput transmission can be achieved by two methods

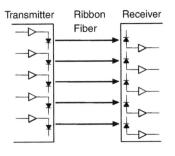

Transmitter Ribbon Receiver
Fiber

Figure 11.2 Concept of a parallel optical link.

of space division multiplexing (SDM) or wavelength division multiplexing (WDM) in optical links. It is thought that SDM optical links that use ribbon fibers are suitable for use over an area of up to several hundred meters. WDM with wavelength combiners and dividers achieves large-throughput, long-distance data transmission that is impossible to provide by serial data transmission or SDM optical links. Because the existing serial fiber can be used, a WDM optical link is suitable for long-distance transmission of several kilometers.

The modules of parallel optical links have the transmitter and the receiver of parallel channels in one package and connect between the two by a ribbon fiber or WDM device. The maximum profit of a parallel optical link is to achieve large-throughput data transmission that suppresses the cost of the module using low-cost optical and electronic parts that operate at a low speed in parallel. A large cooling system with large power consumption is required to achieve high-speed data transmission over 10-Gb/sec with a serial optical data link. At present, a parallel optical link is the main current for large-throughput data transmission.

11.3 DESIGN OF PARALLEL OPTICAL LINK DEVICES

Parallel optical link technology is for an application area connected at a distance from 100 m to several kilometers in LAN and storage-area network (SAN) environments. Parallel optical link technology is suitable for LAN/SAN. For LAN/SAN, the small size, low latency, and large throughput and long distance, low power consumption, ease of use, and cost are issues (Table 11.1).

Because low latency is important for making a data network for a large-throughput connection, it is necessary to suppress the overhead in the delay time caused by O/E data conversion. A low-cost solution is strongly demanded. Parallel optical link technology is the most suitable to meet these demands. Data transmission in the 10-Gb/sec class is achieved in the parallel data transmission system that uses two or more lasers and ribbon fibers or a WDM device.

TABLE 11.1   Requirements for Optical Link Equipped
with Network

	Item	Demand	Remarks
1	Aggregate throughput	1 to 10 Gb/sec	
2	Size of device cable	Both miniaturized	
3	Maximum communication distance	Maximum distance 100 m	Influences the communication method
4	Permissible error rate	Bit error rate $<10^{-20}$	Occurrence of errors not assumed
5	Permissible delay time	Number of clock levels	Much smaller than soft processing
6	Permissible circuit scale	All systems mounted; one board	
7	Permissible cost	Corresponding to existing local-area network (LAN) and storage-area network (SAN)	

Based on these system requirements, it is necessary to execute a design study on parallel optical links for the following items.

1. *Wavelength of laser.* Laser wavelengths are from 850-nm vertical cavity surface-emitting laser (VCSEL) to 1300-nm edge-emitting laser.

2. *The number of parallel optical channels.* The composition of a parallel optical channel varies depending on the total communication throughput and the necessary communication distance. The real implementation also varies depending on the driving integrated circuits (ICs; compound semiconductors, SiGe, BiCMOS, and CMOS).

3. *Coupling method.* There are two methods to achieve alternating current (AC) coupling with scrambling and direct current (DC) coupling. The AC coupling method can achieve high-speed modulation, although the bandwidth is limited. The DC coupling method can achieve wideband from the direct current and is

advantageous for low-latency processing (without encoding and decoding).

4. *Skew compensation method between parallel channels.* Low latency of skew compensation with a small circuit is required.

5. *Module packaging.* Conserving parts to achieve low cost for the optical link module and decreasing the assembly cost are necessary.

6. *Printed circuit board packaging.* When signal processing LSIs and the parallel optical link modules are packaged on the compact printed circuit board, it is necessary to decrease the noise factors of the propagation loss, the signal reflection, cross talk, and so on.

The following sections describe the design manual for these items.

11.3.1 Selection of Laser Wavelength

The parallel optical link achievement method can be performed two major ways by combining the laser and the fiber. These ways are chosen with consideration of data rate, cost, skew, and necessary reliability (error rate) (Figure 11.3).

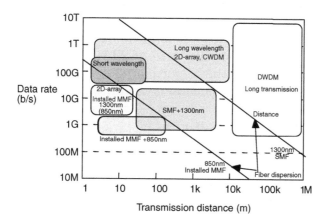

Figure 11.3 Combination of fiber and laser (compared at distance and data speed).

11.3.1.1 Combination of an 850-nm VCSEL and a Multimode Fiber

One method uses an 850-nm wavelength VCSEL for the light source and a multimode fiber (MMF). Because the VCSEL and MMF both have large beam width, the assembly process of the module can be automated. A silicon photodetector for the receiver side provides the low cost.

The skew is large (5 psec/m) because this combination uses a MMF, and skew compensation is needed for a transmission distance of about 100 m at a data rate above 1 Gb/sec. Moreover, as the fiber dispersion is also large (200 MHz·km), a data rate of 1.3 Gb/sec (666 MHz) is the upper limit of the transmission rate when 300-m transmission is assumed.

11.3.1.2 Combination of a 1300-nm Fabry–Perot Laser and a Single-Mode Fiber

A 1300-nm laser wavelength of an edge-emitting (Fabry–Perot-type) laser is used as the light source, and a single-mode fiber (SMF) is used. Because the core diameter of the SMF is small, high accuracy is necessary in the assembly of the module. Because an InGaAs photodetector is used, the modules are expensive.

Because the SMF skew is small (0.5 psec/m), simple and low-latency retiming with the flip-flop can be processed within a distance of about 100 m. Moreover, fiber dispersion is 10 GHz·km, so the data rate of 20 Gb/sec (corresponding to 10 GHz) becomes the upper bound of the modulation rate for an assumed transmission of 1 km.

11.3.2 Selection of Parallel Channel Composition

Fine temperature control is needed to achieve a high-speed (over 10 Gb/sec) transmission rate by a single fiber and single wavelength. Therefore, using parallel optical links with slow data channels (at a data rate of less than 10 Gb/sec) is promising for provision of data transmission above 10 Gb/sec. However, few channels (high-speed driving channels) in the link module are requested.

There are two ways to achieve parallel links: SDM and WDM. The SDM method enhances the method of data transmission with a single channel and makes packaging easier. The WDM method needs a wavelength multiplexer and demultiplexer.

11.3.3 Coupling Method

The technique for making an even appearance of 1 and 0 for a digital signal is necessary for high-speed transmission of a 10-Gb/sec class. For instance, the coding technique called 8B10B is used in the Fiber Channel and Gigabit Ethernet. In addition, the 64B66B coding technique is used in 10-Gigabit Ethernet. The effective transmission rate becomes 8/10 and 64/66 times, respectively, using 8B10B and 64B66B code, and a narrowband signal becomes possible. Moreover, it is easy to achieve a high-speed circuit.

When a signal over 1 Gb/sec is transmitted, the AC coupling type that can guarantee a steady data rate over a narrowband is advantageous. However, the circuit configuration becomes complex for encoding and decoding, and the circuit delay increases.

11.3.4 Skew Compensation Method

There is a difference in the delay time between channels (skew) when synchronous data are transmitted with two or more channels. About 5 ns/m is the propagation delay in the ribbon fiber, and a skew of 500 psec is generated in the 100-m SMF because the characteristic fiber fluctuates by about 0.1%. The average skew here is for a 10-Gb/sec signal transmission; a 100-psec skew is the same as clock cycles. The frame synchronization technique compensates the skew using the synchronous data frame of each channel.

11.3.5 Module Packaging

To achieve a transmission distance of less than several hundred meters (this distance is necessary for connecting a mass interconnection between computing devices in the same floor),

SDM optical links are promising. The main issues in this case are to downsize the device and to reduce the cost of the module with large-throughput, long-distance data transmission (several hundred meters). WDM modules with a transmission distance over several kilometers also require the small-size module packaging.

11.3.5.1 Optical System Packaging

Packaging is more difficult because alignment of submicron accuracy is needed for optical system packaging. Although the core diameter of an MMF is 62.5 or 50 mm, submicron accuracy is necessary to couple the optical fiber of the laser light efficiently. The same fine packaging is similarly necessary for the receiver module. The alignment work for this assembly is a cause of the rising cost of the optical link module. A VCSEL has an advantage from the viewpoint that accurate alignment is easier compared with an edge-emitting laser.

Passive alignment technology solves the problem in packaging. Passive alignment technology does not execute a dynamic optical axis adjustment. Therefore, it reduces the work cost of the alignment compared with active alignment technology (active alignment achieves an optical adjustment by hand or machine adjustment). Passive alignment is increasingly used with technical progress of the VCSEL and large spot size lasers.

11.3.5.2 Electrical System Packaging

For the transmitter in the optical link module, a driver IC is needed for driving the lasers. On the receiver side, amplifiers for slight signals from the photodiodes are needed. It is necessary to package electrical circuits such as drivers and amplifiers compactly and to suppress the influence of parasitic capacitance and inductance for high-speed (10-Gb/sec) drives. Therefore, integrating the high-speed electrical circuit is not easy for packaging in a compact module.

11.3.6 Printed Circuit Board Packaging

When a high-speed optical network is actually developed, high-speed printed circuit board packaging technology will also be needed. In this case, propagation loss, signal reflection, skew, and jitter become problems.

11.3.6.1 Propagation Losses and Signal Reflections

As for the interface for the high-speed circuit, the small amplitude differential signal, such as low-voltage differential signaling (LVDS) or current mode logic (CML), is generally used. The influence of signal disorder on the propagation loss and signal reflection in a printed circuit board is easily received in the signal receiver (the signal amplitude of LVDS is small, from 200 to 400 mV). Therefore, highly accurate circuit impedance control is needed, with a maximum 10-cm wiring distance in the board between components.

11.3.6.2 Skew and Jitter

The arrival time of the received signal varies depending on the characteristic between the parallel signal wires. Suppressing this skew in the synchronous transmission of a high-speed signal becomes a problem. The cycle of a 10-Gb/sec high-speed signal is 100 psec, and the skew that can be allowed is about 5 psec. Therefore, highly accurate characteristic control in the LSI and the I/O buffer of an optical module is needed. Moreover, because the electromagnetic radiation noise that is generated in a high-speed device increases the jitter element of the signal and causes communication errors, jitter becomes a big problem in circuit packaging.

11.4 OPTICAL LINK MODULES

SDM and WDM optical link modules have been developed, and some of them support 10-Gigabit Ethernet, OIF (Optical Internetworking Forum), and InfiniBand standards.

11.4.1 SDM Optical Link Modules

Table 11.2 shows the SDM optical link modules announced by different companies.[10] Many makers have announced various modules. The signals are transmitted by data on the gigahertz

TABLE 11.2 Space Division Multiplexing Optical Link Modules

	Agilent (U.S.) [11]	Alvesta (U.S.) [12]	Aralight (U.S.) [13]	Blaze[a] (U.S.) [14]	Cielo (U.S.) [15]	Corona (U.S.) [16]	Cypress (U.S.) [17]	Emcore (U.S.) [18]	Gore (U.S.) [19]
Lane (b)	12	4	36	12 × 4	12	12	4	12	12
Data rate (Gb/sec)	2.5	3.125	3.125	3.125	2.5	3.35	3.18	2.72	1.6
Total throughput (Gb/sec)	30	12.5	112.5	150	30	40.2	12.72	32.6	19.2
Wavelength (μm)	0.85	0.85	0.85	0.85	1.31	0.85	0.85	0.85	0.85
Length (m)	600	300	300	300	1500	—	100	600	300
Module type	TX/RX	TRX	TX/RX	TX/RX	TX/RX	TX/RX	TRX	TX/RX	TX/RX
Connector	MPO	MPO	MPO, 3 pairs	MPO	MTP	MT	MPO	MPO	MPO

	Infineon (Germany) [20], [21]	NEC (Japan) [22]	NTT (Japan) [23], [24]	Opto-bahn (U.S.) [25]	Paracer (U.S.) [26]	Picolight (U.S.) [27]	Tera-Connect (U.S.) [28]	Xanoptix (U.S.) [29]	Zarlink (Sweden) [30]
Lane (b)	12	4	48	8	12	12	48	36	12
Data rate (Gb/sec)	2.7	3.125	1.25	2.67	2.7	2.7	3.2	2.5	2.5
Total throughput (Gb/sec)	32.4	12.5	60	21.36	32.4	32.4	153.6	90	30
Wavelength (μm)	0.85	0.85	0.85	1300	0.85	0.85	0.85	—	0.85
Length (m)	—	—	—	2000	300	600	300	1100	1500
Module type	TX/RX	TRX	TRX	TRX	TX/RX	TX/RX	TRX	TRX	TX/RX
Connector	MPO	PETIT	MPO	MTP	MTP	MTP	2 × 12MT	6 × 12MT	MPO

Note: TX/RX, transmitter and receiver module; TRX, transceiver module.
[a] 12-channel x 4 lambda CWDM.
Source: From http://www.agilent.com.

order using a VCSEL (850 nm) and 4-, 8-, or 12-channel ribbon fibers. The impedance-matched IC is installed as a special laser driver in the transmitting module and a photodiode driver in the receiving module. The receiver modules use positive-intrinsic-negative (PIN) photodiode arrays. VCSELs are used together with Fabry–Perot-type lasers in the transmission modules. Because monolithic integration with VCSELs is easy and optical coupling efficiency is also higher than for the edge-emitting laser, short-distance connections are mainly used for VCSELs. The main current in the VCSEL wavelength is 850 nm. A modulation rate of 2.5 Gb/sec or more with a VCSEL is possible for each company, and high efficiency in optical coupling to the VCSEL has been achieved using a graded index (GI) 62.5/ 125-μm or 50/125-μm fiber. GI fiber for these models that limits fiber decentralization is 160–500 MHz·km, and the transmission distance is limited to 100 m with a modulation rate of 2.5 Gb/sec.

Three currents for the future technology exist.

1. *High-speed modulation.* The target recognizes over 10 Gb/sec per channel, and operation has already been confirmed for this modulation rate as well as for VCSELs.

2. *Long-distance connection using MMF.* A connection that exceeds several kilometers using installed MMF is requested. There is a possibility that the market can be expanded when connected distances of several kilometers can be achieved using 50/125-μm MMF.

3. *High speed and long distance using SMF.* Long VCSEL wavelengths, such as those suitable for 1.3 and 1.55 μm for SMF, also are in development. Precision assembly to achieve fine optical coupling makes these modules expensive. Cielo (U.S.) has announced development of these types of parallel modules.[15]

In addition to the connector of the multipath optical connector standards (MTP/MPO/MPX) type that have been used so far, the SMC connector is proposed by Infineon.[14]

There are some licenses and movement toward MSA as follows:

PARORI module: Infineon[20,21] and Molex (licensed).[31]

4 + 4 channel transceiver[32]: Agilent, Gore, and Zarlink (MSA).

12-channel transmitting and receiving module[33]: Agilent, Gore, and Zarlink (MSA).

SNAP12[34] (100-pin ballgrid array [BGA] connector and 2.7-Gb/sec/ch optical module): Picolight, Emcore, and Gore (MSA).

Quadlink[35] (four channel transceiver): Alvesta and Cypress (MSA).

11.4.2 WDM Optical Modules

The application of WDM technology for a long-distance optical link was proposed by Blaze.[14] Blaze composes coarse wavelength division multiplexing (CWDM) by 850-nm VCSELs, multiplexes the channel, and has achieved an optical link module with a small size and large capacity (150 Gb/sec). As for WDM technology, there is a big effect from reducing the number of fibers. Moreover, such development is hoped to be a solution to the long-distance difficulties, for which the fiber construction costs are the dominant system costs. It is forecasted that the WDM optical link module class of several kilometers will be proposed by many companies after the development of 1.3-μm wavelength VCSELs.

11.4.3 Integration

The integration of LSI for the parallel link module and signal conditioning enhances the signal speed. In PETIT (the compact parallel OE/EO converter produced by NEC),[22] which mounts the photoelectrical conversion module in the LSI package, a VCSEL and a four-channel photodiode are in the module. The 3.125-Gb/sec transmission speed by one channel has been achieved using an 850-nm VCSEL.

When the data speed becomes 10 Gb/sec, the degradation in the transmission line that ties the LSI and optical module is feared. Furthermore, integration in which both are mounted in the same package (e.g., PETIT) is likely to cause a large reduction of the load to the transmission line.

11.5 10-GIGABIT ETHERNET

11.5.1 Overview

Over a wide area from LAN to WAN, 10-Gigabit Ethernet, which is the latest Ethernet standard, provides easy-to-use interfaces. This standard includes seven optical interfaces and two electrical interfaces.[1] An optical interface has two physical layers (PHY) in a LAN interface, which support the legacy Ethernet frame, and a WAN interface, which supports the SONET/SDH (Synchronous Optical Network/Synchronous Digital Hierarchy) frame. The seven types of optical interfaces were standardized, and in fiscal year 2003 work was completed.[1] Standardization for two types of electrical interfaces (T, CX4) proceeds as a low-cost solution.

The features in an optical 10-Gigabit Ethernet interface are the following:

1. The full-duplex system (carrier sense multiple accesses with collision detection, CSMA/CD) is not used.
2. There are two PHY families (LAN/PHY and WAN/ PHY).
3. There are two types of transmission method (one wavelength serial and four waves wide WDM [WWDM]).

In the data link layer, the Internet Protocol (IP) packet is stored in the frame with the error-detecting function, the transmission data are encoded, and Ethernet is sent to the optical fiber and UTP cable in the PHY (Figure 11.4).

11.5.2 Ethernet Standard Interface

11.5.2.1 LAN/PHY

The PHY in 10-Gigabit Ethernet uses the same data format as the former Ethernet (Table 11.3). Four PHYs can be divided into wide wavelength division multiplexing (WWDM) (LX4), which uses four 1.3-μm wavelengths, and three standards

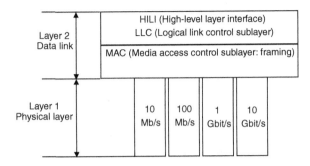

Figure 11.4   Layer structure of standard for 802.3 Ethernet.

(ER, LR, and SR) that use one wavelength. The 1.55-µm semi-conductor laser with an external modulator and SMF are used for ER. A 1.3-µm laser is used for LR. The laser is directly modulated and transmits with SMF. SR can transmit to distances up to 65 m with MMF using a VCSEL that oscillates at 850 nm. After the direct modulation, LX4 is transmitted by one fiber using a laser with four different 1.3-µm wavelengths. It is possible to use both SMFs and MMFs to transmit for

TABLE 11.3   Interface Standards for 10-Gigabit Ethernet and 1-Gigabit Ethernet

			Transmission length		
LAN/PHY	WAN/PHY	Transceiver	Copper	MMF	SMF
10-Gigabits/Second standard					
10GBASE-ER	10GBASE-EW	1.5-µm serial	X	X	40 km
10GBASE-LR	10GBASE-LW	1.3-µm serial	X	X	10 km
10GBASE-LX4	—	1.3-µm WWDM	X	300 m	10 km
10GBASE-SR	10GBASE-SW	0.85-µm serial	X	300 m	
10GBASE-T		Twisted pair	100 m	X	X
10GBASE-CX4		Twisted pair	15 m	X	X
1-Gigabit/Second standard					
1000BASE-LX		1.3-µm serial	X	550 m	5 km
1000BASE-SX		0.85-µm serial	X	220 m	X
1000BASE-T		Twisted pair	100 m	X	X

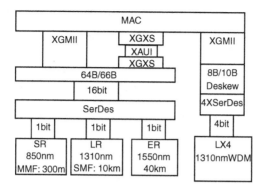

Figure 11.5   Layer structure of LAN/PHY for 10-Gigabit Ethernet.

distances of up to 30 m with a MMF. FDDI grade MMF (62.5/125-µm type) can be used. If the latest wideband MMF of the 50/125-µm type is used, data can be transmitted up to 300 m.

The functional structure of LAN/PHY is described in Figure 11.5. In the serial PHY ER, LR, and SR, a 2-b header is applied to every 64 b (64B/66B). 64B/66B is the encoding method used to convert the transmission data from a media access control (MAC) layer into the serial DC-balanced data. The conversion efficiency of the 64B/66B code has improved to 97% (=64/66) compared with the efficiency of 80% in 8B/10B code used in the past Ethernet standard. When using 64B/66B and a physical data transfer velocity, some speeds increase from 10 Gb/sec to 10.3125 Gb/sec. The 64B/66B code adds a synchronous 2-b header every 64 b that scrambles; the DC balance is improved (when it searches for the two synchronization bits on the receiver side and the head of the scramble code is specified).

In LX4, 8B/10B is encoded, 4 bytes of serial data are made separately, and the bit string of 4 × 3.125 Gb/sec is converted into four different wavelengths and transmitted. A mechanism that compensates the skew between the four wavelengths is installed, and parallel synchronous transmission of the four lanes is achieved. LX4 is the electricity of each lane that can slow the modulation rate on a channel by using a parallel, synchronous transmission of the four lanes. Low-price components can be used for electrical conversion. The serial signal of the four lanes in LX4 is the same interface as the electrical interface 10 Gigabit Attachment Unit Interface (XAUI). The serial interface XAUI has a parallel interface 10G Media Independent Interface (XGMII) (7 cm connected distance) and the same functional block composition as the PHY block of WWDM.

11.5.2.2 WAN/PHY

Connection with an existing optical transmission device becomes possible with the PHY input and output with the SONET/SDH frame. This frame accommodates the serial bit string that performs the block encoding in OC-192/VC-4-64c. The transmission throughput is 9.29 Gb/sec, which is considerable (to insert the idle pattern beforehand in the MAC sublayer, the frame rate is suppressed from 10 Gb/sec). The standards of the PHY include three standards (EW, LW, and SW) of the same serial systems as LAN/PHY. Although now deleted, the standard of WWDM existed at the beginning.

A suitable structure for connecting the SONET/SDH infrastructure exists so that WAN/PHY can use an interchangeable signal speed and framework with SONET. WAN/PHY in XGMII and XAUI has a common layer structure with LAN/PHY (Figure 11.6). The WAN interface sublayer (WIS) is installed in WAN/PHY and accommodates the serial bit string that encodes 64B/66B in the SONET/SDH frame. It helps the SONET frame make a synchronous payload for 64B/66B code that reaches from a higher layer, adds the transport overhead, and transmits. The layer from serializer and deserializer (SerDes) to the optical transmitting-and-receiving system uses LAN/PHY and three common serial transmission systems (EW, LW, SW).

11.5.2.3 Electrical Cable Interface

Anxiety about high costs is widespread, and two standards (10GBASE-T, 10GBASE-CX4) that use electrical cable are newly advocated. 10GBASE-T uses UTP cable and supports

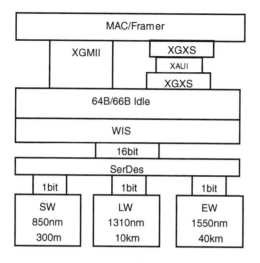

Figure 11.6 Layer structure of the WAN/PHY for 10-Gigabit Ethernet.

a wiring distance of up to 100 m. UTP cable has interchangeability with 10BASE-T, 100BASE-T, and 1000BASE-T. 10GBASE-CX4 makes wiring within 15 m a target by using shielded Cu cable.

11.5.3 MSA

MSA at the part level, such as optical transceiver activity, is advancing with standardization of the interface for Ethernet.[36] MSA is expected to encourage the expansion of market opportunity because it will help provide the second and third choices to users without changing the shape or the pin layout of the part. It is thought that MSA will become more active in the future.

11.5.3.1 XENPAK

XENPAK[37] is the MSA for the transceiver module for 10-Gigabit Ethernet that was proposed by Agilent and Agere. The XAUI interface is installed, and the signal is input and output with a card-edge-type electrical connector with 70 pins (supporting hot plug). An optical interface supports four types of standards:

850-nm serial, 1310-nm WWDM, 1310-nm serial, and 1550-nm serial supported by 10-Gigabit Ethernet. The interface inputs and outputs the signal with the two-ream SC connector. The module is 51.3 mm wide, 22.4 mm high, and 121 mm long. The module enlarged the encoding sublayer where XGXS and 64B/66B were done by adopting the XAUI interface compared with GBIC and so on. On the other hand, the I/O signal line was reduced using the XAUI interface compared with SFI-4.

11.5.3.2 XPAK

XPAK[38] is the MSA for the module for 10-Gigabit Ethernet that was advocated by Infineon and Picolight. XPAK supports two interfaces (XAUI and SFI-4-P2) and can correspond to three applications: 10-Gigabit Ethernet, 10-Gb Fiber Channel, and SONET. The module is the size of a small, round dice (from 36 mm wide, 9.8 mm high, and 75.69 mm long) with XENPAK modules. An optical interface has four interfaces: optical serial, parallel light, electrical cable, and WDM. The data communication speed supports everything from 9.95 to 11.2 Gb/sec.

11.5.3.3 XFP

XFP[39] is the MSA for the module for 10-Gigabit Ethernet that aims at small-form factor (SFF). Broadcom (U.S.), Brocade (U.S.), Emulex (U.S.), Finisar (U.S.), JDS Uniphase (U.S.), Maxim (U.S.), ONI Systems (U.S.), Sumitomo Electric Industries (Japan), Tyco Electronics (U.S.), and Velio (U.S.) are participating in development. OC-192/STM-64, 10-Gb Fiber Channel, and 100 Gigabit Ethernet are supported. The speed of inputting and outputting the signal by a 10-Gb/sec serial electrical signal is different from XENPAK and XPAK. XFP is hot pluggable, and the size is about a third that of XENPAK (18.35 mm wide, 8.5 mm high, and 62.1 mm long). The electrical signal interface adopts the XFI standard. A distance of up to 12 in. can be connected with a printed circuit board of FR4 material, and XFI connects in the 500-mV differential motion serial signal line (it is a pair terminal at 100 Ω).

11.5.4 Institute of Electrical and Electronics Engineers 802.3 Standard

11.5.4.1 A Standardization Organization

The Institute of Electrical and Electronics Engineers (IEEE) is an international society, and standards are developed by the IEEE Standard Association (IEEE-SA), which belongs to IEEE (Figure 11.7). Some standards committees exist in IEEE-SA. The standard committee related to Ethernet was established in February 1980. The committee name is based on the date of establishment; that is, IEEE 802 LAN/MAN SC is the local-area network and metropolitan-area network standards committee. This is sometimes abbreviated to LMSC. LMSC coordinates with other standards groups. There is strong international participation, and some meetings are held outside the United States. LMSC is separated into some working groups (WGs), and they deliberate on standardization.[40] Now, there are 14 WGs as well as 2 disbanded WGs. Of these groups, 8 are inactive, and the other 6 groups are working (i.e., 802.1, 802.3, 802.11, 802.15, 802.16, and 802.17). Two technical advisory groups (TAGs; 802.18 and 802.19) were inaugurated at the plenary meeting in July 2002. The role of the TAGs is to assist the WGs. Standards of Ethernet for optical or copper cables are developed by IEEE 802.3 WG.[41] IEEE 802.3 WG is a WG that discusses signal transmission technology of carrier sense multiple accesses with collision detection (CSMA/CD). CSMA/CD technology was not used with 10-Gigabit Ethernet.

There are three plenary IEEE 802.3 SA meetings every year, and interim meetings are held as needed. There is no

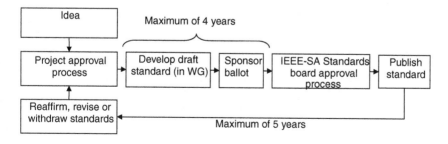

Figure 11.7 The outline of a developing standards process.

restriction to participate in these meetings, and participants can qualify to ballot by attending three meetings. Those qualified to ballot are called *voters* and have the right to vote on a resolution. The resolutions are determined if 50% of all voters cast their votes, and support for the resolution can be made by obtaining 75% of all ballots. The outline of the developing standards process is shown in Figure 11.7. It is necessary to develop standards in 4 years; once standards are enacted, they cannot be reaffirmed, revised, or withdrawn for 5 years.

11.5.4.2 Hierarchy of IEEE 802 from the Viewpoint of IEEE 802.3 WG

Figure 11.8 shows the hierarchy of each WG, mainly focusing on IEEE 802.3; this expresses the Open System Interconnection (OSI) model. The IEEE 802.1 WG develops standards for the network layer (protocol layers above the MAC and logical link control [LLC] layers). The IEEE 802.2 WG develops standards for the LLC sublayer. This working group is currently

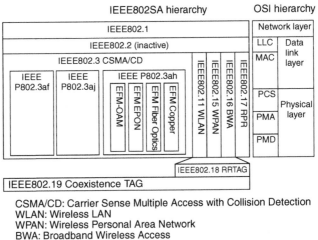

Figure 11.8 The hierarchy of each WG standardization procedure.

inactive, with no ongoing projects. IEEE 802.3 WG develops standards for the MAC sublayer and the PHY. The PHY consists of the physical coding sublayer (PCS), which encodes the transmission channel; the physical media attachment (PMA) sublayer for direct-parallel conversion; and the physical-media-dependent (PMD) sublayer, which defines the transceiver that depends on the media. For example, the PMD for 10-Gigabit Ethernet has standards such as the 850-nm, 1310-nm, and 1550-nm lasers.

11.5.4.3 The Types Published in IEEE 802.3

Table 11.4 shows the list of Ethernet types published in IEEE 802.3. Each underline shows a widely used type. The last number in the type (e.g., the 4 of 10GBASE-LX4) expresses the number of parallelisms of the cables. There are two transmission methods; one is BASE, which is a baseband transmission method. The other is BROAD, which is a broadband transmission method. The BROAD method has 10BROAD36, which supports a 3.6-km connection length. In addition, fiber-optic interrepeater link (FOIRL) connects repeaters spaced up to 1 km and forms the basis of 10BASE-F. 10BASE-F means 10BASE-FP, 10BASE-FB, and 10BASE-FL collectively. 100BASE-T means 100BASE-TX, 100BASE-T2, and 100BASE-T4 collectively. 1000BASE-X means 1000BASE-LX, 1000BASE-SX, and 1000BASE-CX collectively. These transmission methods are also known as the task force (TF). Table 11.5 shows the TF and type names. IEEE 802.3ae was published at the first plenary meeting in 2002.

11.5.4.4 Current Situation of the IEEE 802.3 WG

The following TFs were active in the plenary meeting in March 2003:

P802.3af, DTE Power via MDI. WG is called Power Over Ethernet (PoE).

P802.3ah, Ethernet in the First Mile (EFM). WG concerns the first connection between a customer and the Internet communication carrier.

TABLE 11.4 Ethernet Types and Notation of BASE

Category	Code	Description	Examples
Distance	5	500 m	1 BASE5, 10BASE5
	2	185 m	10BASE2
Transmission media	T	Twisted pair	10BASE-T, 100BASE-TX,-T4, -T2, 100BASE-T2,
	F	Fiber	10BASE-FP, -FB,-FL, 100BASE-
	C	Coax	1000BASE-CX
Type of laser (Fiber)	E	Extra long (40 km, 1550 nm)	10GBASE-ER,-EW
	L	Long (10 km, 1310 nm)	10GBASE-LR,-LW,-LX4, 1000BASE-LX
	S	Short (65 m, 850 nm)	10GBASE-SR,-SW, 1000BASE-
Transmission rate	1		1BASE5, etc.
	10	10 Mb	10BASE-FP, etc.
	100	100 Mb	100BASE-T4, etc.
	1000	1000 Mb (1 Gb)	1000BASE-CX, etc.
	10G	10 Gb	10GBASE-LX4, etc.

TABLE 11.5 Task Force and Type

Task force	Year published	Type
802.3	1983	10BASE5
802.3a	1985	10BASE2
802.3b	1985	10BROAD36
802.3d	1987	FOIRL
802.3e	1987	1BASE5
802.3i	1990	10BASE-T
802.3j	1993	10BASE-FP, 10BASE-FB, 10BASE-FL
802.3u	1995	100BASE-TX, 100BASE-FX, 100BASE-T4
802.3y	1997	100BASE-T2
802.3z	1998	1000BASE-LX, 1000BASE-SX, 1000BASE-CX
802.3ab	1999	1000BASE-T
802.3ae	2002	10GBASE-LX4,10GBASE-SR, 10GBASE-LR, 10GBIT/SE-ER, 10GBASE-SW, 10GBASE-LW,

P802.3aj, Maintenance TF. WG concerns published contents.
P802.3ak, 10GBASE-CX4 TF. WG concerns 10-Gigabit
Ethernet using four coaxial cables. 10GBASE-CX4 is
supported under 15 m and attains low cost, in contrast
to the expensive optical 10-Gigabit Ethernet. 10GBASE-
CX4 is applicable to an external backplane connection
or an aggregator for stackable switches and interserver
connection in a data center. It is expected to expand the
10-Gigabit Ethernet market.

The following are subtask forces (STFs) in P802.3ah:

EFM-OAM (Operations, Administration, and Maintenance). STF concerns maintenance management.
Ethernet–Passive Optical Network (EFM EPON). STF
concerns point-to-multipoint communication by the
optical fiber.
EFM Fiber Optics. STF concerns optical fiber.
EFM Copper. STF concerns copper wiring, and the present
main subjects are ADSL related.

At the plenary meeting held in November 2002, the following Call for Interests and new SGs were established:

10GBASE-T. Transmission distance exceeding 100 m is supported, using a Category 5, 5e, or 6 or other twisted-pair media. Associated undertakings apply to transmission feasible on about 25 m, and some have a rosy forecast for 100-m transmission. If 10GBASE-T is attained, 10-Gigabit Ethernet will spread rapidly because of its low cost.

Security. In point-to-multipoint communication, security issues are noted because broadcast is the top down communication.

A 10-Gigabit Ethernet consortium was established at the University of New Hampshire.[42]

11.5.4.5 Anticipated Future of IEEE 802.3 WG Trends

The next stage of 10-Gigabit Ethernet is attracting attention. However, there is no present Call for Interest, and there is no motion regarding post–10-Gigabit Ethernet. There are currently two opinions whether the post–10-Gigabit Ethernet should be 40 Gb or 100 Gb. It is thought that post–10-Gigabit Ethernet will be 100 Gb because the Ethernet standard has progressed at a times ten development rate. It is expected that a Call for Interest will be announced by the end of 2005, and the standard of post–10-Gigabit Ethernet will be published around 2007 or 2008. A new coding technology for 10-Gigabit Ethernet, 64B/66B, was developed to enable application of the optical module used by SONET/SDH and to realize a variable-length frame and a 10-Gb full rate. Thus, to know the trends of post–10-Gigabit Ethernet, it is necessary to observe the developing situation of a device or circumference technology.

11.6 OPTICAL INTERNETWORKING FORUM

The Optical Internetworking Forum (OIF)[2] is advocated as a standard technology that connects the network products of the switch and the router. The SxI OIF standard with a high-speed, optical interface connects the SONET/SDH framer and SerDes (x is the standard F [Framer] or P [packet]). The Very Short Reach-x (VSR-x, for which x varies from 1 to 4) standards, which are short-distance link interfaces, have already reached agreement.

11.6.1 SxI Standard

The SxI standard concerns the electrical interfaces between the SONET framer and SerDes. They are classified into several groups (OC, optical carrier [SONET terminology]; STM, synchronous transport module [SDH terminology]):

SxI-3 OC-48/STM-16 (2.48832 Gb/sec range)
SxI-4 OC-192/STM-64 (10 Gb/sec range)
SxI-5 OC-768/STM-256 (40 Gb/sec range)

11.6.2 VSR Standard

The VSR standard is advocated as an optical interface for an intraoffice network in OC-192 of OIF. It has four categories that use serial interfaces together with parallel interfaces (Table 11.6).

11.6.2.1 VSR-1

VSR-1 (Figure 11.9) uses the same laser and MMF used in Gigabit Ethernet. It forwards data at a data rate of 1.25 Gb/sec per channel using an 850-nm, 12-channel VCSEL array and 12-channel ribbon fiber. The 62.5/125-μm fiber (bandwidth of 400 MHz·km) is used; communication can occur at distances

TABLE 11.6   Standards for Very Short Reach (VSR) in OC-192

Name	Transmission distance	Fiber type	Number of fibers	Laser type and wavelength
VSR-1	300 m	MMF	12	VCSEL/850 nm
VSR-2	600 m	SMF	1	FP/1310 nm
VSR-3	300 m	MMF	4	VCSEL/850 nm
VSR-4	300 m	MMF	1	VCSEL/850 nm

Note: FP, Fabry–Perot laser.

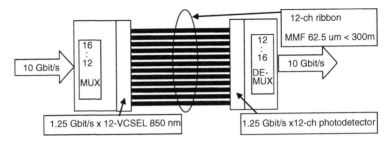

Figure 11.9 Structure of VSR-1.

up to 300 m. The fiber connector uses the MPO/MTP type. Ten fibers transmit data, one fiber communicates data for cyclic redundancy check (CRC) error detection, and one communicates the parity of ten data signals.

11.6.2.2 VSR-2

In VSR-2 (Figure 11.10), communication is for a distance of up to 600 m using one laser (1310-nm wavelength) and SMF. An uncooled 1310-nm Fabry–Perot-type laser in accordance with ITU G.691 is used (to decrease device cost).

11.6.2.3 VSR-3

In VSR-3 (Figure 11.11), 850-nm VCSEL and MMF are used similar to VSR-1. Two-way communications are achieved using one 12-channel ribbon fiber. Four fibers transmit at a data rate of 2.5 Gb/sec, and another four fibers transmit in the opposite direction. The 50/125-μm fiber (500-MHz·km bandwidth) is

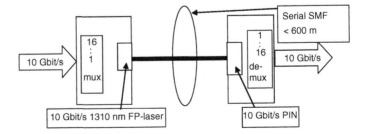

Figure 11.10 Structure of VSR-2.

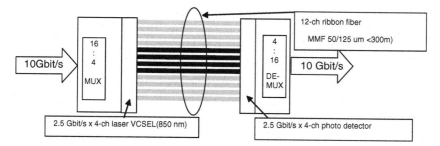

Figure 11.11   Structure of VSR-3.

used; communication can take place at distances up to 300 m. The fiber connector is the MPO/MTP type.

11.6.2.4 VSR-4

Two-way 10-Gb/sec communication is achieved using MMF and 850-nm VCSEL (Figure 11.12). When the 50/125-μm fiber (500 MHz·km bandwith) is used, communication is at distances of up to 85 m. When the 2000-MHz·km bandwidth fiber is used, communication is at distances of up to 300 m.

11.6.3   MSA

The following discussion is of MSA activities related to OIF.

11.6.3.1   300-Pin MSA

The 300-pin MSA is the standard for a 10-Gb/sec optical transponder with an electrical connector with 300 pins. An

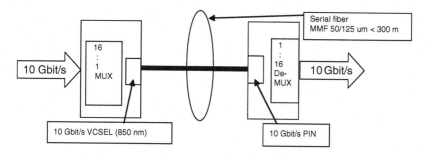

Figure 11.12   Structure of VSR-4.

electrical interface conforms to SFI-4, and an optical interface is in accordance with ITU-T corresponding to a transmission distance of up to 65 km. The light source uses a long-wavelength 1.3- or 1.55-µm laser. To install the high-power laser with an external modulator and to use the cooling element for the temperature stability, it includes a large-scale heat sink. An electrical connector uses the Meg-Array connector with 10 rows of 30 pins (FCI). A 300-pin SFF transponder standard was decided as an optical transponder for VSR. The positions of the connector shape and the screw hole are standard and can be substituted on the printed circuit board for a MSA for 300 large-scale pins.

11.6.3.2 200-Pin MSA

The 200-pin MSA[36] (ToMCat: 10-Gb module coolerless transceiver) is the standard for a small 10-Gb/sec optical transponder. Power consumption is limited to 7.5 W, and miniaturization of the external size is achieved. This is used for an optical transponder for short-distance transmission at the 2- to 10-km level.

11.7 INFINIBAND

InfiniBand[3] is expected to be the I/O standard for the next-generation server. In InfiniBand, a parallel, large-throughput connection of 2.5, 10, and 30 Gb/sec is achieved with 1, 4, and 12 as the units of the 2.5-Gb/sec signal line. It aims at communication among computer nodes, for which large throughput and low latency are important. The standard is classified into SX, used to connect for VSR, and LX, to connect for longer distances (Table 11.7). In InfiniBand, a 2.0-Gb/sec electrical signal modulation rate is commonly encoded, and the optical signal is commonly communicated at the speed of the transmission of the 2.5-Gb/sec signal.

11.7.1 1X-SX and 1X-LX

Two serial fibers interactively connect 1X wide (Figure 11.13). It is divided into the 1X-SX and 1X-LX standards, depending on the transmission distance. Three specifications (laser wavelength, fiber, and coding) are different. The 1X-LX link

TABLE 11.7   InfiniBand Standard

	Very short reach (VSR)	Longer reach
1X Wide		
Name	IB-1X-SX	IB-1X-LX
Wavelength	850 nm	1300 nm
Connector	Dual LC	Dual LC
Distance	250 m (with 50/125 μm MMF)	10 km (SMF)
	125 m (with 62.5/125 μm with MMF)	
4X Wide		
Name	IB-4X-SX	N/A
Wavelength	850 nm	
Connector	Single MPO	
Distance	125 m (with 50/125 μm MMF)	
	75 m (with 62.5/125 μm with MMF)	
12X Wide		
Name	IB-12X-SX	N/A
Wavelength	850 nm	
Connector	Dual MPO	
Distance	125 m (with 50/125 μm MMF)	
	75 m (with 62.5/125 μm with MMF)	

uses MMF, and 1X-SX uses SMF. The LC-type connector is used.

11.7.2   4X-SX

4X-SX (Figure 11.14) achieves four interactive connections with eight fibers. Generally, ribbon fiber is used. MMF is used,

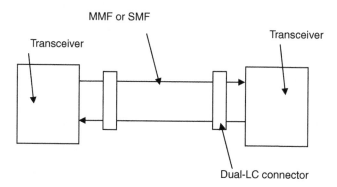

Figure 11.13   InfiniBand 1× wide. MMF, multimode fiber; SMF, single-mode fiber.

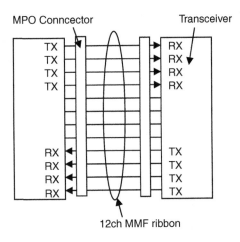

Figure 11.14   InfiniBand 4× wide.

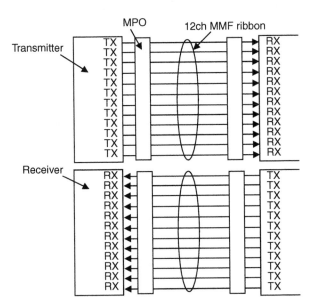

Figure 11.15   InfiniBand 12× wide.

as is the MPO connector. When 12-channel ribbon fibers are used, the four fibers at the center are not used.

11.7.3 12X-SX

12X-SX (Figure 11.15) achieves an interactive connection in 12 systems by using 24 fibers (two ribbon fibers of 12 channels). The MPO connector of two reams is used.

REFERENCES

1. IEEE 802: http://www.ieee802.org/3/ae/index.html.

2. Optical Internet Forum: http://www.oiforum.com.

3. Infiniband: http://www.infinibandta.org/home.

4. J.W. Goodman, F.I. Leonberger, S.Y. Athale, and R.A. Kung, Optical interconnects for VLSI system, *Proc. IEEE*, 72, 159 (1984).

5. D.A.B. Miller and H.W. Ozaktas, Limit to the bit-rate capacity of electrical interconnection from the aspect ratio of the system architecture, *J. Parallel Distributed Comput.*, 41, 42 (1997).

6. S. Nishimura, H. Inoue, H. Matsuoka, and T. Yokota, Optical interconnection subsystem used in the RWC-1 massively parallel computer, *IEEE J. Sel. Top. Quantum Electron.*, 5, 360 (1999).

7. S. Nishimura, T. Kudoh, H. Nishi, K. Harasawa, N. Matsudaira, S. Akutsu, and H. Amano, 64-Gb/sec highly reliable network switch (RHiNET-2/SW) using parallel optical interconnection, *IEEE J. Lightwave Tech.*, special issue on Optical Networks, 18, 1620 (2000).

8. S. Nishimura, T. Kudoh, H. Nishi, J. Yamamoto, R. Ueno, K. Harasawa, S. Fukuda, Y. Shikichi, S. Akutsu, K. Tasho, and H. Amano, RHiNET-3/SW: an 80-Gb/sec high-speed network switch for distributed parallel computing, *Hot Interconnect*, 9, 119 (2001).

9. A. Takai, T. Kato, Y. Yamashita, S. Hanatani, Y. Motegi, K. Ito, H. Abe, and H. Kodera, 200-Mb/sec/ch 100 m optical subsystem interconnection using 8-channel 1.3-μm laser diode array and single-mode fiber arrays, *IEEE J. Lightwave Technol.*, 12, 260 (1994).

10. I Schmale: http://www.paralleloptics.org/.

11. Agilent: http://www.agilent.com.

12. Alvesta: http://www.alvesta.com.

13. Aralight: http://www.aralight.com.

14. Blaze: http://www.blazenp.com.

15. Cielo: http://www.cieloinc.com.

16. Corona: http://www.coronasys.com.

17. Cypress: http://www.cypress.com.

18. Emcore: http://www.emcore.com.

19. Gore: http://www.goreelectronics.com/.

20. Infineon: http://www.infineon.com/.

21. D. Kuhl, K. Drogemuller, J. Blank, M. Ehlert, T. Kraeker, H. Hohn, D. Klix, V. Plicker, L. Melchior, P. Hildebrandt, M. Heinemann, A. Beier, L. Leininger, H.D. Wolf, T. Wipiejewski, and R. Engel, PAROLI® a parallel optical link with 15 Gb/sec throughput in a 12-channel wide interconnection, *Sixth Int. Conf. Parallel Interconnects*, 187 (1999).

22. T. Yoshikawa, I. Hatakeyama, K. Miyosi, and K. Kurata, Optical Interconnection as an intellectual property of a CMOS library, *Hot Interconnect*, 9, pp. 124–129, (2001).

23. A. Ohki, N. Sato, M. Usui, N. Tanaka, K. Katsura, T. Kagawa, and Y. Ando, Development of 60 Gb/sec-class Parallel Optical Interconnection Module (ParaBIT-1F), IPSJ SIGNotes computer Architecture, Abstract 138-007, Japan, 2001.

24. K. Katsura, Y. Ando, Y. Usui, A. Ohki, N. Sato, N. Matsuura, N. Tanaka, T. Kagawa, and M. Hikita, Para-BIT: parallel optical interconnection for large-capacity ATM switching system, *Trans. Elec. Comm.* E82-B, 412 (1999).

25. Optobahn: http://www.optobahn.com.

26. Paracer: http://www.paracer.com.

27. Picolight: http://www.picolight.com.

28. Teraconnect: http://www.teraconnect.com.

29. Xanoptics: http://www.xanopptics.com.

30. Zarlink Semiconductor: http://www.zarlink.com.

31. Molex: http://www.molex.com.

32. Agilent: http://ftp.agilent.com/pub/semiconductor/morpheus/docs/44_pfm_spec_1dot0.pdf.

33. Agilent: http://ftp.agilent.com/pub/semiconductor/morpheus/docs/12x_pfm_spec_1dot0.pdf.

34. SNA: http://www.snapoptics.org/.

35. Alvesta: http://www.alvesta.com/quadlink.html.

36. 10 Giga Multisource Agreement: http://www.10gigmsa.com/.

37. Xenpak: http://www.xenpak.org/.

38. XPAK: http://www.xpak.org/.

39. XFP: http://www.xfpmsa.org/.

40. IEEE 802: www.ieee802.org.

41. IEEE 802: www.ieee802.org/3.

42. University of New Hampshire Research Computing Center: www.iol.unh.edu.

43. 300 PIN MSA http://www.300pinmsa.org/.

12

Broadband Networks and Optical Interconnections

SEIJI KOIZUMI

CONTENTS

12.1 TRANSMISSION SYSTEMS

12.1.1 Broadband Network System

Optical transmission systems provide the lowest loss and highest transmission rate for the broadband network in comparison with other systems. The broadband network is expanding because of the worldwide popularization of the Internet. Broadband networks are only for data communication, not traditional voice communication like the PSTN (public switched telephone network), and involves exchange of digital information between a computer and other digital devices using packet switching. The broadband network system is composed of the local-area network (LAN; office or home), access network (fiber to the home, FTTH), metropolitan network (MAN), wide-area network (WAN; long haul or submarine cable). Figure 12.1 shows an overview of broadband networks.

LAN: The LAN interconnects users in a local area, such as a home, a building, an office or factory, a campus, and the like.

Access network: The access network interconnects the LAN and an access point of the metro edge.

MAN: The MAN interconnects a narrow area, such as several building blocks or an entire city and metropolitan area.

WAN: The WAN interconnects neighboring cities, long-distance cities, or across a country.

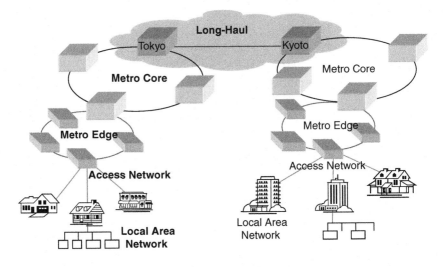

Figure 12.1   Broadband networks.

12.1.2   Broadband Network Basics

12.1.2.1 Network Elements

To understand a digital communications network, the network is usually conceptualized. A network is composed of only three elements: terminal, link, and node. A *terminal* is an end machine, such as data equipment or a telephone. A *node* is a point joining another node or terminal. A *link* is a line connecting nodes or a node and a terminal. The simplest network is from terminal, to link, to node, to link, to terminal. Figure 12.2 shows the network model. This model is useful for understanding network structure.

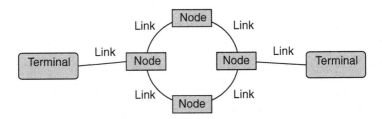

Figure 12.2   Network model.

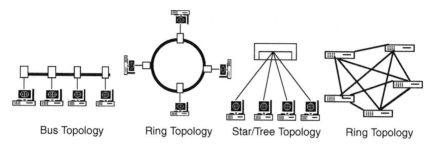

| Bus Topology | Ring Topology | Star/Tree Topology | Ring Topology |

Figure 12.3 Network topology.

12.1.2.2 Network Topology

Topology shows the physical and logical mapping of the network. Figure 12.3 shows four typical topology types. Each has its own particular advantages and limitations in terms of expandability, reliability, performance, and equipment and link cost. In practical networks, these topologies are combined for optimizing performance and cost.

12.1.2.3 Bus Topology

All of the network nodes are connected to a common transmission line and share a medium. Only one pair of terminals on the bus can transmit data, and other terminals cannot send data at the same time. All stations can see all data and can enter data only using their own address.

12.1.2.4 Ring Topology

Information in the form of data packets travels around the ring from node to node. All the nodes have input and output ports, which are always active, and transfer a packet to the next node. When the node only recognizes its own address, the node accepts data and does not transfer data to the next node.

12.1.2.5 Star/Tree Topology

In the star/tree topology, all the nodes are connected to a single port, called a *hub* or *concentrator*. All data moving from node to node always pass through the hub port. The hub can see all data and handle it.

12.1.2.6 Mesh Topology

In mesh topology, all nodes are connected to each other by point-to-point links. Nodes have many ports and link lines. The mesh topology provides a flexible and reliable network, but the cost is high. This is used in a PSTN voice network.

12.1.3 Open System Interconnection Seven-Layer Reference Model

The International Organization for Standardization created the Open Systems Interconnection (OSI) reference model to reduce incompatibility between computers. Figure 12.4 shows the OSI model; this model has seven layers that define the tasks that must be performed to transfer information on the network. It is only a conceptual framework for understanding network structure. From physical layer (PHY) to network layer, it is important to understand the broadband system.

Layer 1: The PHY defines how a medium connects to a computer device, electrical signal, optical connector, optical fiber, and so on.

Layer 2: The data link layer places groups of data in containers. Data are a 1 or 0 and are packaged into one frame. The frame (bits) includes synchronization, error control, and flow control.

Layer 3: The network layer identifies routing, how to transfer information over the network.

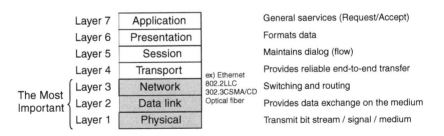

Figure 12.4   OSI seven-layer model.

Layer 4: The transport layer corrects transmission errors and provides end-to-end error recovery and flow control capability.

Layer 5, 6, 7: Application area.

12.1.4 Optical Transmission Link

Broadband network systems transmit a digital signal using IM/DD (intensity modulation/direct detection). Figure 12.5 shows a diagram of an IM/DD link.

The electrical input signal data stream to the optical transmitter is nonreturn-to-zero (NRZ) digital. The electrical signal is converted to an optical on/off signal by laser or light-emitting diode (LED). This optical signal is propagated through fiber cable and to the optical receiver. The optical signal is then converted back into the electrical signal by a photodiode, such as an Avalanche photodetector (APD).

12.1.5 Wavelength Division Multiplexing System

Figure 12.5 shows a single-wavelength digital data transmission. For more effective use of optical fiber, which has a high potential transmission capability, and limitation of data rate speed, a time domain multiplexing (TDM) transmission system is used. Figure 12.6 shows a wavelength division multiplexing (WDM) system. Optical sources are combined in the multiplexer to a single fiber. Erbium-doped fiber amplifiers (EDFAs) are used to amplify all the wavelength channels simultaneously. An add/drop multiplexer adds or drops a wavelength signal.

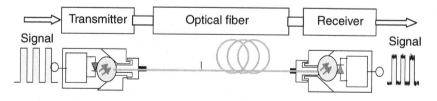

Figure 12.5 Intensity modulation/direct detection (IM/DD) transmitter/receiver (Tx/Rx).

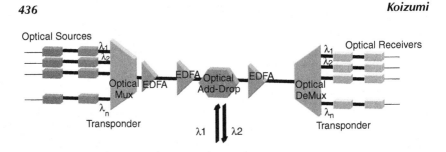

Figure 12.6 WDM system. DeMux, demultiplexer; Mux, multi-plexer.

12.1.6 Internet Protocol Packet

12.1.6.1 Packet Frame

In the broadband Internet Protocol (IP) network, a datum is divided into appropriate size containers for bits termed packet-like Ethernet frames or IP frames. The packets are transferred and reach the destination address for reassembly of the bits into the datum. All packets have a header that includes a destination address, a source address, and data. Figure 12.7 shows the packet frame segmentation-and-reassembly process.

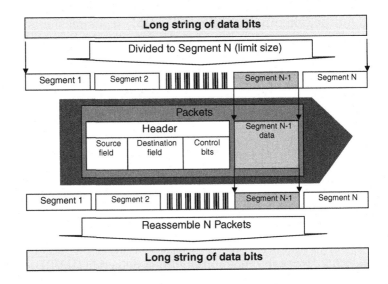

Figure 12.7 Segmentation-and-reassembly process.

12.1.6.2 Packet Switching

Packets are forwarded through the network in a series of intermediate nodes termed routers. A node transfers only the proper neighboring node in a manner similar to a relay in which a bucket is passed from one person to another. A packet is transferred node to node individually and finally arrives at the destination address.

12.2 ETHERNET USING OPTICAL LINKS

12.2.1 Overview

In the early 1970s, Metcalfe at Xerox Palo Alto Research Center put forward the Ethernet LAN. At the same time, other key Internet technologies were announced: Transmission Control Protocol/Internet Protocol (TCP/IP) by Cerf and a 4-b microprocessor from Intel. These three inventions were initiated in California. In the 1980s, Ethernet was deployed as the DIX (DEC Intel Xerox) Ethernet standard. In 1982, Ethernet was submitted to the Institute of Electrical and Electronics Engineers (IEEE) as the 802.3 document and became the IEEE standard for LAN. Ethernet has become the most widely used protocol for LANs and is extending to the MAN or WAN. The IEEE 802.3 Standard Committee covers all new technologies, such as wireless LAN or 10-Gigabit Ethernet.

12.2.2 IEEE 802.3 Family

The IEEE 802.3 committee provides the Ethernet family of specifications (Table 12.1). The first part of the name (10, 100, 1000) indicates the speed in megabits per second. The second part (BASE) means the baseband transmission mode. The last portion is a distance indicator or a PHY designation.

12.2.3 Carrier Sense Multiple Access
with Collision Detection

Ethernet uses an access method called the Carrier Sense Multiple Access with Collision Detection (CSMA/CD) Protocol. The CSMA/CD Protocol determines which terminals get the network and when they get to use it. Only one terminal is permitted to

TABLE 12.1 Ethernet Standard Family

	Name	Medium	Max L	Remarks
Ethernet: 10 Mb/sec speed	10BASE5	50-Ω coaxial	500 m	
	10BASE2	50-Ω coaxial	185 m	
	10BASE-T	Category 3 UTP cable	100 m	
	10BASE-FP	MMF/850 nm	1 km	Passive Star
	10BASE-FB	MMF/850 nm	2 km	Backbone
	10BASE-FL	MMF/850 nm	2 km	Link
Fast Ethernet: 100 Mb/sec speed	100BASE-TX	Category 5 UTP	100 m	
	100BASE-T4	Category 3 UTP	100 m	
	100BASE-FX	MMF	2 km	2 pair fiber
Gigabit Ethernet: 1 Gb/sec speed	1000BASE-T	Category 5e/6 UTP (IEEE 802.3ab)	100 m	
	1000BASE-CX	Shielded twisted pair	25 m	Duplex
	1000BASE-SX	MMF/850 nm (IEEE 802.3z)	550 m	Duplex
	1000BASE-LX	MMF/1310 nm (IEEE 802.3z)	550 m	Duplex
		SMF/1310 nm (IEEE 802.3z)	5 km	
10-Gigabit Ethernet: 10 Gb/sec speed 802.3ae	10GBASE-LX4	SMF/1310 nm/WDM	10 km	For LAN
	10GBASE-SR	MMF@BW500M/1310 nm/WDM	300 m	For WAN
	10GBASE-LR	MMF@BW500M/850 nm	82 m	
	10GBASE-ER	SMF/1310 nm	10 km	
	10GBASE-SW	SMF/1550 nm	40 km	
	10GBASE-LW	MMF@BW500M/850 nm	82 m	
	10GBASE-EW	SMF/1310 nm	10 km	
		SMF/850 nm	40 km	

Note: MMF, multimode fiber; SMF, single-mode fiber; UTP, unshielded twisted-pair.

talk at any time. If the network is in use by one terminal, other terminals should wait until the network is silent. The terminal always listens to the network. If two terminals start to use the network at the same time, a collision will occur, ending data transmission. Each terminal waits a specified time and starts resending.

12.2.4   Logical Link Control Protocol

The Logical Link Control (LLC) Protocol layer resides above the media access control (MAC) layer. The LLC layer uses an unacknowledged, connectionless service and depends on a best-effort mode of frame delivery. The data unit in LLC is the protocol data unit (PDU), which contains (variably) a destination service access point (DSAP), a source service access point (SSAP), and control and information parts. Figure 12.8 shows the field format of the LLC Protocol PDU.

12.2.5   MAC Frame

Ethernet's MAC layer is responsible for formatting MAC frames by the LLC layer above it. The frame contains synchronization (preamble), flag, flow, and error control information. Figure 12.9 is an example of a MAC frame in IEEE 802.3. Each field has the following function:

Preamble: The 8-b square-wave pattern 10101010 is repeated seven times for all receivers preparing a signal and obtaining synchronization.

Start frame delimiter (SFD): The sequence is fixed as 10101011, and receivers find the first bit of the frame.

Destination address: The destination address is the physical address of the next destination.

DSAP 8 bits	SSAP 8 bits	Control 8 bits or 16 bits	Information variable bits

Figure 12.8   Field format of the LLC Protocol data unit.

Preamble	SFD	Destination address	Source address	Information LLC data	Pad	FCS
7 bytes	1	6 bytes	6 bytes	◄——— 46-1500 bytes ———►		4 bytes

Figure 12.9 Example of Ethernet media access control (MAC) frame of IEEE 802.3. LLC, logical link control.

Source address: The physical address of the last device that forwarded the packets.

Length: Length indicates the number of bytes in the PDU.

Information: Information is the entire data from the LLC.

Pad: The pad adjusts the frame byte length; a minimum of 64,000 bytes is required for collision detection.

Frame check sequence (FCS): The FCS has error detection information for the frame and is based on a 32-b cyclic redundancy check (CRC).

12.2.6   Gigabit Ethernet

Gigabit Ethernet is specified as high-speed 1-Gb/sec LAN in IEEE 802.1z. Gigabit Ethernet is compatible with 10BASE-T and 100BASE-T in frame structure and procedure. The PHY differs in the encoder and decoder; the autonegotiation and flow control functions are specified as the physical coding sublayer (PCS). Figure 12.10 shows Gigabit Ethernet architecture in IEEE 802.3z and 802.3ab.

12.2.7   10-Gigabit Ethernet

The IEEE 802.3ae standard for 10-Gigabit Ethernet was released in 2002. This standard established an Ethernet with a transmission speed of 10 Gb/sec. The previous Ethernets were designed for LANs. This specification defines two types of PHY: the LAN PHY and the WAN PHY. These PHYs define the physical layer for entire MANs, WANs, and LANs.

IEEE 802.3ae uses the IEEE 802.3 Ethernet MAC Protocol, frame format, and frame size. 10-Gigabit Ethernet provides only full-duplex operation and does not support shared-medium implementations. Therefore, it logically has no distance limitation. It uses only an optical fiber link with a

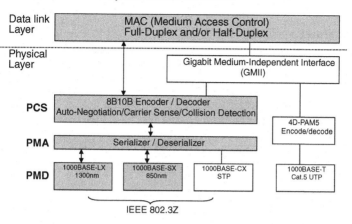

IEEE 802.3Z

Figure 12.10 Gigabit Ethernet architecture in IEEE 802.3z. PCS, physical coding sublayer; PMA, physical media attachment; PMD, physical media dependent; STP, shielded twisted pair; UTP, unshielded twisted pair.

distance from 65 m to 40 km by single-mode fiber (SMF) and multimode fiber (MMF) with 850, 1310, and 1550 nm wavelength. The specification covers various applications, from long haul to LAN. The WAN PHY option allows transparency between OC-192 Synchronous Optical Network (SONET), not SONET interfaces; it is an asynchronous Ethernet interface.

The following hold for high-speed applications with 10-Gigabit Ethernet:

- High-speed Internet access for ease of Internet environment use.
- Enterprise LAN interconnections to provide high-speed intranet.
- Real-time streaming for contents delivery service.
- Metro link to provide low-cost equipment and seamless connection.

12.3 TRANSMISSION MEASUREMENT

Fiber-optic transmission system measurements determine if the system meets its design goal. There are three major methods to analyze the physical characteristic of an optical system: bit error rate test (BERT), waveform analysis, and jitter test.

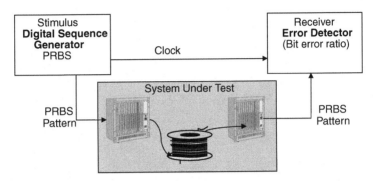

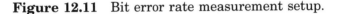

Figure 12.11   Bit error rate measurement setup.

12.3.1   Bit Error Rate Measurement

The most important measurement in a transmission system is the rate at which errors occur end to end. A common evaluation method is the BERT. Figure 12.11 shows the BERT setup. A stimulus signal pattern is injected into the system, equipment, and fiber cable. A pseudorandom binary sequence (PRBS) is often used to simulate a wide range of bit patterns. The PRBS sequence is a random sequence of bits and repeats a random bit stream pattern. An error detector compares the bit stream from the system tested and the known bit stream and counts the errors. A common measurement is of the bit error rate as a function of loss of transmission line.

$$\text{BER} = \frac{\text{Number of Bits Received in Error}}{\text{Number of Bits Received}}$$
$$= \frac{\text{Error Count in Measurement Period}}{(\text{Bit Rate}) \times (\text{Measurement Period})}$$

12.3.2   Waveform or Eye Measurement

The BERT indicates whether a transmission line is good. To find the cause of error, it is necessary to analyze the received waveform and find the margin in performance. A high-speed oscilloscope is used to observe digital waveforms at the output port as an eye diagram or eye pattern. Figure 12.12 shows an eye diagram example.

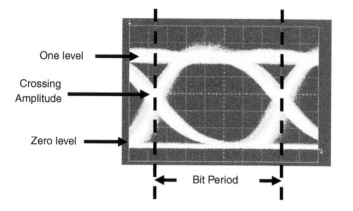

One level

Crossing
Amplitude

Zero level

Bit Period

Figure 12.12 Eye pattern.

12.3.3 Jitter Measurement

Jitter is defined as short-term phase variation of the significant instants of a digital signal from its ideal positions in time. A most easy measurement is to use an oscilloscope, but it has several limitations (for example, measurement range and sensitivity cannot provide spectral characteristics and jitter time function).

12.4 LOCAL-AREA NETWORK BY OPTICAL LINK

12.4.1 Local-Area Network

The LAN is used for an office computer network, high-speed localized backbone, wireless LAN, storage-area network (SAN), home network, campus network, and so on. A LAN is usually used, owned, and operated by a single organization. To increase the distance and capacity (speed), LANs use optical fiber links.

12.4.2 LAN Switch

For increasing LAN capacity significantly and constructing a larger network than can be achieved with a central hub, an Ethernet switched solution is available. The advantage of switched Ethernet is that it eliminates the collision detection timing limitation, extends the distance limitation between stations, and allows the simultaneous access of multiple input/output (I/O) ports. It means no sharing of line and bandwidth. Each access port has a buffer memory and queues incoming frames while waiting for the outgoing port to become available. In 100BASE-FX, a full-duplex connection can switch the input/output ports individually and simultaneously.

12.4.3 Enterprise Router

Enterprise routers connect to multiple dedicated-leased communication lines named as WANs to enable LAN users to access the Internet and connect to remote offices. Most enterprise networks are built from Ethernet segments by Ethernet switches or bridges and are connected to outside sites. Routers connect a LAN to the outside world through high-speed, broadband digital lines. Enterprise routers have interfaces for both high-speed Ethernet LANs and high-speed WANs. They switch LAN/WAN interface ports and connect other internal segments or external networks by referring the network layer's packet destination address. The router has two basic functions: determining optimal paths from the source address to the destination address and transporting packets through the network.

12.5 ACCESS NETWORK USING OPTICAL LINKS

12.5.1 Overview

Optical transmission systems for trunk lines such as WAN and MAN, metro-core, long-haul, or international submarine cable systems are used worldwide. Optical transmission systems are now spreading to broadband access networks to the home or buildings and are called the last 1 mile or first 1 mile.

12.5.2 FTTx

FTTH using only optical fiber is the best access solution in broadband applications. Considering fiber line cost, media are mixed; optical fiber and metal cable are named FTTx.

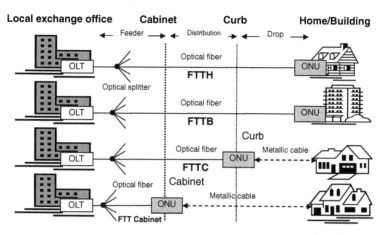

Figure 12.13 Classification of FTTx. FTTB, fiber to the building; FTTC, fiber to the curb; OLT, optical line terminal; ONU, optical network unit.

Figure 12.13 shows the classification of FTTx. For example, FTTC is a solution in which fiber is run close to the home and to the curb and metallic cable runs from the curb to the home (ADS, active double star). In place of metallic cables, coaxial cables for cable antenna television (CATV) and wireless systems for fixed wireless access (FWA) are used. As a final or ideal solution, FTTH brings fiber into the residence.

12.5.3 FTTH Optical Network Topology

When compared with metallic cable, fiber has excellent characteristics, such as bandwidth and low loss. Fiber material is expensive. Many network topologies are proposed that consider characteristics and cost (Figure 12.14). The passive double star (PDS) system saves optical fiber length with a simple passive optical device.

Single star (SS): The network directly connects the user or home to the telephone office without optical splitters or electrical multiplex equipment. This is a simple system.

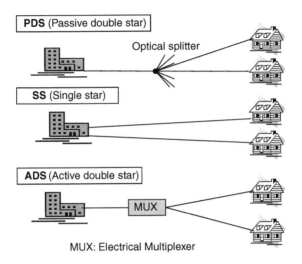

Figure 12.14   Network topology.

PDS: The PDS is a passive optical network (PON) using passive components such as an optical splitter (using a fused-fiber coupler) or a WDM component (using a waveguide-type interferometer). It needs no power supply in the network node.

ADS: The ADS uses an active electrical component for time multiplexing.

12.5.4   Bidirectional Transmission Method

Two fibers for a single channel are called space division multiplexing (SDM) and provide full-duplex connection easily. To decrease fiber cable cost, bidirectional transmission using one fiber cable is necessary to connect between the telephone office and the home. WDM, time compression multiplexing (TCM), and code division multiplexing (CDM) are proposed. In WDM, upstream and downstream transmission links are made over a single fiber at different wavelengths. TCM uses a burst signal that is time compressed and switches the upstream and downstream bursts at a specified time interval. CDM uses unique codes for both streams. The optical signal is modulated by a unique Code A and transmitted. The receiver demodulates the

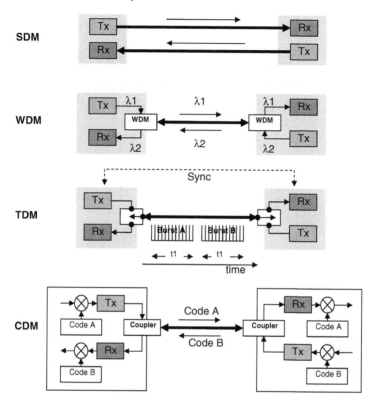

Figure 12.15   Bidirectional transmission method. Rx, receiver; Tx, transmitter; SDM, space division multiplexing; TDM, time division multiplexing.

optical signal using the same unique Code A. Figure 12.15 shows SDM, WDM, TDM, and CDM bidirectional transmission links.

12.6   PASSIVE OPTICAL NETWORK

A passive optical network (PON) allows multiple users to share a single optical fiber line without using electrical equipment at halfway to users. This method reduces not only the cost in fiber line, but also the cost of fiber construction and the scalability of user channel and bandwidth. Figure 12.16 is an overview of broadband PON (B-PON) and Ethernet PON (E-PON) systems.

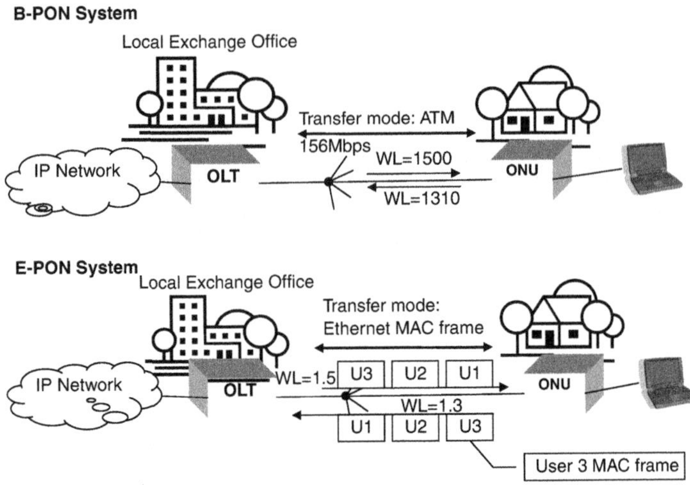

B-PON System

E-PON System

Figure 12.16   PON systems.

12.6.1   Broadband PON

The B-PON standard, formerly ATM-PON, was formed by the FSAN (Full Service Access Networks Initiative) Coalition in 1995. In 1999, the ITU-T had approved the G.983.1 and G.983.2 specifications. The asynchronous transfer mode (ATM) is used upstream at 156 and 622 Mb/sec and downstream at 156 Mb/sec using an ATM cell.

12.6.2   Ethernet PON

In 2004, the E-PON standard was created as IEEE 802.3ah, EFM (Ethernet in the First Mile) by the IEEE 802.3 Subcommittee. The frame will use an Ethernet MAC frame, and the maximum rate is 1 Gb/sec. The maximum distance will be up to 10 km using SMF, and a 16-line split is expected.

12.7   MEDIA CONVERTER

A media converter (MC) is used for extending the distance between personal computer (PC) stations or networks to other stations or networks. MCs are beginning to be used for FTTH in a SS topology by a 100-Mb/sec or 1-Gb/sec line.

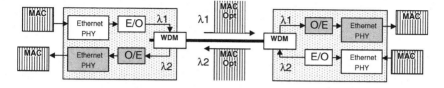

Figure 12.17 Media converter block diagram. E/O, electrical to optical; O/E, optical to electrical.

Ethernet frames are transferred and only convert in the PHY. Figure 12.17 shows a MC using WDM. The IEEE 80.2.3ah standard was created in 2003.

12.8 METROPOLITAN AND LONG-HAUL NETWORK

12.8.1 Synchronous Digital Hierarchy/SONET

Synchronous Digital Hierarchy (SDH; ITU standard)/SONET (American National Standards Institute [ANSI] standard) is the first common standard that applies to networks worldwide in high-speed networks using optical fiber for WANs. The SDH/SONET specifications define a frame format, physical interfaces, and strong network management tools; as a result, SDH/SONET provides a reliable network and quick recovery time. Table 12.2 shows the signal hierarchy. Many SDH/ SONET

TABLE 12.2 SDH/SONET Signal Hierarchy Structure

OC Level	SDH	SONET	Data rate (Mb/sec)	Payload rate (Mb/sec)
OC-1	STM-0	STS-1	51.48	50.840
OC-3	STM-1	STS-3	155.52	150.336
OC-9	STM-3	STS-9	466.56	451.008
OC-12	STM-4	STS-12	622.08	601.344
OC-18	STM-6	STS-18	933.12	902.016
OC-24	STM-8	STS-24	1244.60	1202.688
OC-36	STM-12	STS-36	1866.00	1804.032
OC-48	STM-16	STS-48	2488.32	2405.376
OC-96	STM-32	STS-96	4876.64	4810.752
OC-192	STM-64	STS-192	9953.28	9621.504

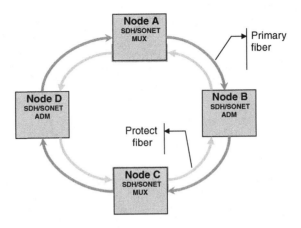

Figure 12.18 SDH/SONET.

systems are used in metropolitan and long-haul networks because they have high speed, reliability, and compatibility for the existing PSTN. Almost all SDH/SONET networks are configured by a ring topology using a dial counter rotating ring with one pair of optical fibers for survivability and quick recovery. Figure 12.18 shows SDH/SONET ring architecture. Even IEEE 802.3ae 10-Gigabit Ethernet WAN PHY has connectivity with SDH/SONET interfaces.

12.8.2   Dense WDM

Long-haul or international optical fiber networks are increasing long-distance transmission capacity using dense WDM (DWDM) and EDFA technologies. Also, DWDM is under exploration for MANs for increasing channel capacity.

12.8.3   Routers

To transport Internet data in metropolitan or long-haul networks, most traffic uses IP packets and switches to point to point or point to multipoint. To connect to the Internet, a local network must send and retrieve data using TCP/IP and related protocols. Networks also connect to other sites, known as intranets or extranets, using IP. Routers and switches are the key equipment. Routers work at Layer 3 of the OSI model and switch and route packets across multiple networks. Routers read and examine network addressing information and the destination address in the IP header and then send the packet to the next routers by routing information shared with neighboring routers. The packet route is a hop-by-hop routing model similar to a bucket brigade relay. In MANs or WANs, routers have many high-speed physical interfaces, Ethernet, SDH/SONET, ATM, ISDN, and so on. A terabit switch router has been developed and belongs to an emerging class of backbone platforms offering terabit capacity and supporting interfaces that range from OC-3 (155 M) to OC-192 (10 G).

12.8.4   Resident Packet Ring

Resident packet ring (RPR) is the new technology for the metropolitan ring and is similar to an SDH/SONET ring (see Figure 12.18). It uses Ethernet MAC technologies. The IEEE 802 Committee created the 802.17 standard. The RPR will provide a reliable network for high-speed Ethernet base networks in the same manner as the SDH/SONET ring.

12.9   NEXT-GENERATION ALL-OPTICAL NETWORK

12.9.1   Overview

In the optical networking revolution, networks will channel all traffic over a single fiber connection and will provide redundancy using a mesh interlocking pathway. If a line breaks, traffic will change the pathway within a second using optical switches. The next-generation optical networks will use an end-to-end optical architecture, meaning that nowhere in the network is the optical signal converted into an electrical signal. Figure 12.6 shows the all-optical network using DWDM mux/demux (multiplexer/demultiplexer), EDFA, and optical add/drop Mux.

12.9.2   Optical Add/Drop Multiplexer

In next-generation system network should be easily reconfigurable in real time without electronics. Optical add/drop

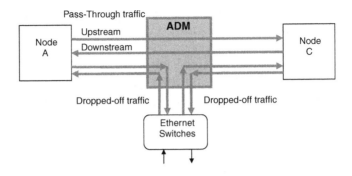

Figure 12.19   Optical add/drop multiplexer (OADM).

multiplexers (OADMs) use special filters to select the wavelengths that channels need to drop off. Figure 12.19 shows OADM signal flow. This provides a specific wavelength to a specific customer, meaning customers are able to purchase or lease wavelengths.

12.9.3   Optical Cross Connect

An optical cross connect (OXC) is located at a junction point in optical networks and provides end-to-end connection using a wavelength path. OXCs link any of several incoming wavelength lines to any of several outgoing lines and reroute traffic when a network fails. An OXC provides dynamic wavelength management, reconfiguration, and automated optical layer provisioning. Figure 12.20 shows an OXC.

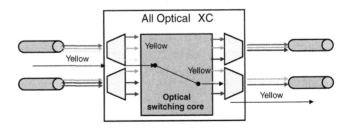

Figure 12.20   Optical cross connect (OXC).

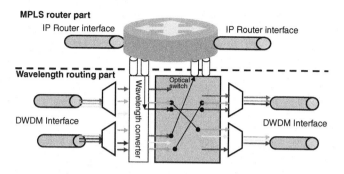

Figure 12.21 Multiprotocol labeling switch (MPLS) network. DWDM, dense wavelength division multiplexing; IP, Internet Protocol.

12.9.4 Optical IP Routing

To provide more flexible and dynamic optical IP networks, the MPLS (multiprotocol labeling switch) is in development. Using MPLS, optical IP routers switch the optical path by examining the IP header. Figure 12.21 shows an example of an MPLS router.

12.10 SUMMARY

This chapter presented an introduction to LAN and WAN network basic technology and focused on Ethernet and FTTH. Ethernet has become the most widely-used network system, not only LAN area. And it is growing to access network and metropolitan network area. Now in Ethernet, the transmission speed became up to 10GHz. For understanding FTTH (Fiber to the home) systems, both Ethernet and Optical transmission is very important technology. The Broadband network is expanding to personal home areas using fiber optic and Ethernet technology. It is not so far to realize the full fiber optical network.

Index